Analyzing High-Dimensional Gene Expression and DNA Methylation Data with R

T0136303

Hongmei Zhang

CRC Press
Taylor & Francis Group
Boca Raton London New York

CRC Press is an imprint of the
Taylor & Francis Group, an **informa** business
A CHAPMAN & HALL BOOK

First edition published 2020
by CRC Press
6000 Broken Sound Parkway NW, Suite 300, Boca Raton, FL 33487-2742

and by CRC Press
2 Park Square, Milton Park, Abingdon, Oxon, OX14 4RN

Library of Congress Cataloging-in-Publication Data

Names: Zhang, Hongmei (Professor of Biostatistics), author.
Title: Analyzing high-dimensional gene expression and DNA methylation data with R / Hongmei Zhang.
Description: Boca Raton, FL : CRC Press, 2020. | Series: Chapman & Hall/CRC mathematical and computational biology series | Includes bibliographical references and index.
Identifiers: LCCN 2020006652 (print) | LCCN 2020006653 (ebook) | ISBN 9780367495169 (paperback) | ISBN 9781498772594 (hardback) | ISBN 9780429155192 (ebook)
Subjects: LCSH: Gene expression--Statistical methods. | Gene expression--Data processing. | DNA--Methylation. | Epigenetics--Statistical methods. | R (Computer program language)
Classification: LCC QH450 .Z43 2020 (print) | LCC QH450 (ebook) | DDC 572.8/65--dc23
LC record available at https://lccn.loc.gov/2020006652
LC ebook record available at https://lccn.loc.gov/2020006653

ISBN: 9781498772594 (hbk)
ISBN: 9780367495169 (pbk)
ISBN: 9780429155192 (ebk)

Typeset in LM Roman 12
by Nova Techset Private Limited, Bengaluru & Chennai, India

Visit the Taylor & Francis Web site at
http://www.taylorandfrancis.com

and the CRC Press Web site at
http://www.crcpress.com

Analyzing High-Dimensional Gene Expression and DNA Methylation Data with R

Chapman & Hall/CRC Mathematical and Computational Biology

About the Series

This series aims to capture new developments and summarize what is known over the entire spectrum of mathematical and computational biology and medicine. It seeks to encourage the integration of mathematical, statistical, and computational methods into biology by publishing a broad range of textbooks, reference works, and handbooks. The titles included in the series are meant to appeal to students, researchers, and professionals in the mathematical, statistical, and computational sciences and fundamental biology and bioengineering, as well as interdisciplinary researchers involved in the field. The inclusion of concrete examples and applications and programming techniques and examples is highly encouraged.

Series Editors

Xihong Lin
Mona Singh
N. F. Britton
Anna Tramontano
Maria Victoria Schneider
Nicola Mulder

Computational Exome and Genome Analysis
Peter N. Robinson, Rosario Michael Piro, Marten Jager

Gene Expression Studies Using Affymetrix Microarrays
Hinrich Gohlmann, Willem Talloen

Big Data in Omics and Imaging
Association Analysis
Momiao Xiong

Introduction to Proteins
Structure, Function, and Motion, Second Edition
Amit Kessel, Nir Ben-Tal

Big Data in Omics and Imaging
Integrated Analysis and Causal Inference
Momiao Xiong

Computational Blood Cell Mechanics
Road Towards Models and Biomedical Applications
Ivan Cimrak, Iveta Jancigova

An Introduction to Systems Biology
Design Principles of Biological Circuits, Second Edition
Uri Alon

Computational Biology
A Statistical Mechanics Perspective, Second Edition
Ralf Blossey

Computational Systems Biology Approaches in Cancer Research
Inna Kuperstein and Emmanuel Barillot

Introduction to Bioinformatics with R
A Practical Guide for Biologists
Edward Curry

Analyzing High-Dimensional Gene Expression and DNA Methylation Data with R
Hongmei Zhang

For more information about this series please visit: https://www.crcpress.com/Chapman--HallCRC-Mathematical-and-Computational-Biology/book-series/CHMTHCOMBIO

To my family

Contents

Preface

The writing of this book is motivated by my research in the area of genetics and epigenetics and a course that I have been teaching recently with a focus on biostatistics methods used to analyze bioinformatics data. The rapidly advancing "omics" technologies have generated massive transcriptomics, metabolomics, proteomics, and epigenomics data and brought a strong potential to promote biomedical research. Raw "omics" data are filled with noise and biased by batches and technical variations, and furthermore, this data covers tens of thousands of genes which have complex relationships and interconnections. Students majoring in Biostatistics and researchers dealing with these types of data have a strong need to learn methods and analytical tools that have the ability to handle these high-dimensional data starting from the initial product, the raw data. These skills include but are not limited to data preprocessing, data mining, marker detection, and other related analyses. However, books having a "pipeline" of analytical methods with concrete examples are rather limited.

The book is an attempt to fill this gap and aims to provide a "pipeline" for gene expression and epigenetic data analysis starting from raw genome- and epigenome-scale data. For epigenetic data, we specifically focus on DNA methylation. The book is a combination of methodology introduction and R program implementation. Unlike many R programming books which focus on utilizing R packages, this book introduces each statistical method before moving to concrete examples and R packages. This structure is expected to substantially benefit biostatisticians as well as readers with interest in the underlying techniques in addition to implementation of R packages. A number of analytical methods and tools are discussed in the book, including methods to preprocess genome-scale gene expression and epigenetic data, methods for data mining to identify potentially informative factors, and methods for subsequent analyses after data mining, e.g., factor/variable selections, network construction, and testing for differential networks. Many methods introduced are recently developed and advanced, which are expected to outperform existing ones. In the meantime, these advanced approaches will provide a platform for statisticians to develop more efficient approaches. For most methods introduced, they are followed by examples and R programs for users to practice and apply.

This book covers statistical methods and specific examples with R programs. It does not require strong statistical background to use the programs, and thus has a potential to apply to readers with different needs. The targeted readers of this book are graduate students with some statistics/biostatistics background and biostatisticians with interest in analyzing high-dimensional "omics" data, especially expressions of genes and DNA methylation, and researchers interested in analyzing "omics" data with limited knowledge in statistical methods. For graduate students, it is ideal if they are in the second year of their study. For PhD students, there is no limitation. Students will have a better understanding of the statistical methods and programs if they have some statistic background. For researchers, they are not required to have strong statistic backgrounds, but some understanding in statistic concepts will be very helpful.

In the process of writing this book, I was supported by a number of people. Especially, I would like to thank David Grubbs and his assistants for their patience with me and their flexibility to fit my academic schedule. The support from the School of Public Health at the University of Memphis made the whole writing process pleasant and productive. I am thankful to Luhang Han, Shengtong Han, Yu Jiang, and Meredith Ray for their help in proof-reading. My great appreciations also go to my husband and children for their understanding and generosity. I would like to dedicate this book to my parents, who are always supportive.

## CHAPTER 1

Introduction

1.1 PIPELINES TO ANALYZE "OMICS" DATA

High throughput technologies allow researchers to monitor in parallel expression levels of thousands of genes or strength in DNA methylation at a vast number of CpG sites. However, raw "omics" data are usually filled with noise. For instance, both Illumina Infinium HumanMethylation450 BeadChip (450K methylation array) and the Infinium MethylationEPIC BeadChip (850K methylation array) have different probe types, each using different chemistry. The process of bisulphite conversion of DNA, chip to chip variation, and other steps introduce assay variability and batch effects. Thus before we start analyzing the data, quality control and data preprocessing are needed. Commonly applied pipelines in general all start from quality control and preprocessing (Figure 1.1). Sometimes preprocessing is omitted if doing so removes biological relevance.

In this book, the focus is on genome-scale gene expression data and DNA methylation data. Quality control and data preprocessing are introduced and discussed in Chapters 2 and 3, along with a brief review of the data generating process. Due to the high dimensionality of data, it is necessary to exclude uninformative loci, i.e., loci potentially not associated with the study of interest. Doing so will dramatically improve statistical power in the detection of associations or identification of biological markers. However, we have to admit that screening can end up with false negatives, that is, excluding loci which are actually informative. The methods and examples for screening are discussed in Chapter 5. Which covers methods that utilize training and testing data, surrogate variables, and an approach of sure independence screening. All these approaches are built upon regressions.

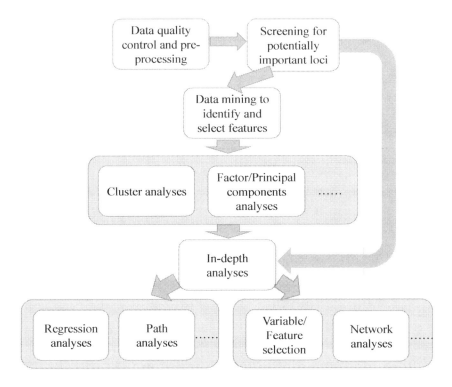

Figure 1.1 An example of the pipeline to analyze "omics" data. The step of preprocessing can be omitted if this step potentially removes biological relevance.

Data mining is an important step to identify underlying features and patterns in the data. Findings from this step will substantially benefit subsequent in-depth analyses. Commonly used data mining techniques include cluster analyses, factor analyses, and principal component analyses. We focus on cluster analyses in this book. Compared to factor and principal component analyses, results from cluster analyses allow researchers to concretely visualize profiles of each cluster and identify the uniqueness of each cluster. In addition, interpretation of results from cluster analyses is relatively straightforward. Cluster analyses are discussed in Chapter 5, including classical approaches such as partitioning-based methods and hierarchical clustering approaches and joint clustering methods where the clustering is two-dimensional.

Often, after screening or data mining, in-depth analyses are conducted. Standard approaches such as linear regressions or generalized linear regressions are commonly used. However, even after screening, the number of loci (or variables) can still be large. Thus, efficiently detecting markers of an exposure or for a health outcome is critical in medical research. To this end, accompanied by concrete examples, this book focuses more on methods to detect markers or select important variables. Chapters 6 and 7 introduce variable selection techniques in linear and non-linear models, and Chapter 8 discusses methods and examples for network constructions and comparisons. In some studies, DNA methylation is treated as a mediator between an exposure and a phenotype of interest. In this case, mediation analyses via path analyses can be applied. This is not covered by this book but recent studies have proposed methods to assess such mediation effects.

Through out the book, we utilize simulated data as well as real gene expression and DNA methylation data to demonstrate the analytical methods. All the programs in this book have been tested in R version 3.5.3 and/or 3.6.1. In the following sections of this chapter, we briefly introduce each real data set.

1.2 RNA-Seq GENE EXPRESSION IN S2-DRSC CELLS

Brooks et al. [13] conducted a study aiming to explore the conservation of the splicing code between distantly related organisms, in particular, between *Drosophila* and mammals. To identify regulatory targets of Pasilla, S2-DRSC cells were treated with a 444 bp dsRNA fragment corresponding to the *ps* mRNA sequence. Untreated S2-DRSC were cultured in biological triplicate to serve as a control. The authors combined RNAi and RNA-Seq to identify exons regulated by Pasilla, the *Drosophila melanogaster* ortholog of mammalian NOVA1 and NOVA2. The RNA-Seq data for the treated and untreated cells and related information are available from the Gene Expression Omnibus (GEO) database under accession numbers GSM461176-GSM461181. The data are available at `https://figshare.com/s/e08e71c42f118dbe8be6`. The reads were aligned to the *Drosophila* reference genome and summarized at the gene level. This RNA-Seq data set is utilized to demonstrate the methods and packages for clustering in Chapter 5.

1.3 MICROARRAY GENE EXPRESSION IN YEAST CELLS AND IN PROSTATE SAMPLES

Zhao et al. [163] examined three microarray gene expression data sets across the yeast cell cycle and identified 254 genes that are periodic in at least two data sets. Expressions of genes with periodicity were further examined in the study by [127, 114] to demonstrate their developed methods and R packages for clustering. They extracted expression data of 256 genes collected in the first 16 time points with 7-minute-intervals. As noted in Qin et al. [114], expressions of these 256 genes are cell cycle dependent. A subset of the data with 64 genes is analyzed in Chapter 5.

We also examined another microarray gene expression data set discussed in Singh et al. [129]. The raw data set has Affymetrix expressions of 52 tumoral and 50 non-tumoral prostate samples. A set of preprocessing steps were applied, including setting thresholds at 10 and 16, 000 units, excluding genes with expression variation less than 5-fold relatively or less than 500 units absolutely between samples, applying a base 10 log-transformation, and standardizing each experiment to zero mean and unit variance across the genes. In the package depthTools, expressions of 100 genes were included, representing the most variables genes in expression as noted in Dudoit et al. [30]. The variation was measured as a ratio of between-group to within-group sum of squares in expression of genes. This data set is discussed in Chapter 8.

1.4 DNA METHYLATION IN NORMAL AND COLON/RECTAL ADENOCARCINOMA SAMPLES

DNA methylation of 38 matched pairs is available in the Cancer Genome Atlas (TCGA) data repository. Among the 76 samples, 38 have colon and rectal adenocarcinoma. This data set is available in the DMRcate package used to identified differentially methylated regions. In this book, we utilize this data set to demonstrate methods in marker detections and variable/feature selections (Chapters 6 and 7).

## CHAPTER 2

Genome-scale gene expression data

For genetic data, we mainly focus on gene expression data produced via Sanger sequencing and next generation sequencing. For epigenetic data, which will be discussed in the next chapter, we focus on DNA methylation of CpG sites. Single nucleotide polymorphisms in genomewide association studies will not be discussed.

2.1 MICROARRAY GENE EXPRESSION DATA

2.1.1 Data generation

Different techniques are available to measure genome-scale gene expressions, for example, cDNA spotted arrays and oligonucleotide arrays (Figure 2.1). For cDNA spotted array, it is the first type of DNA microarray technology developed in the Brown and Botstein Labs at the Stanford University [14]. It was produced by using a robotic device, which deposits a library of thousands of distinct cDNA clones onto a coated microscope glass slide surface in serial order with a distance of approximately 200-250 μm from each other, one spot for one gene. These moderate sized glass cDNA microarrays also bear about 10,000 spots or more on an area of 3.6 cm^2. Then mRNA samples or targets from two groups (e.g., treatment and control samples) were extracted, separately reverse transcribed into cDNA, and labeled with different fluorescent dyes (e.g., red color Cyanine-5 or Cy5 and green color Cyanine-3 or Cy3). The mixture of these labeled cDNA were hybridized onto the microarray, competing to bind to the cDNA probes. After hybridization, the slides are imaged using a scanner or a charge-coupled device camera to obtain

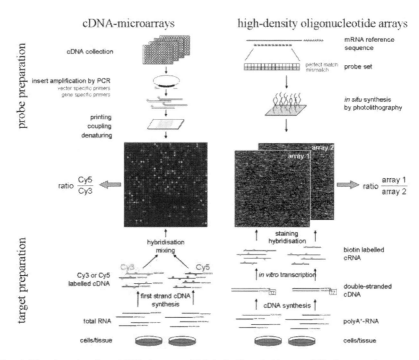

Figure 1. Schematic overview of spotted cDNA microarrays and high-density oligonucleotide arrays. **cDNA microarrays**: Array preparation: inserts from cDNA collections or libraries are amplified and the PCR products printed at specified sites on glass slides using high-precision arraying robots. These probes are attached by chemical linkers. Target preparation: RNA from 2 different tissues or cell populations is used to synthesize cDNA in the presence of nucleotides labeled with 2 different fluorescent dyes (eg: Cy3 and Cy5). Both samples are mixed in a small volume of hybridization buffer and hybridized to the array, resulting in competitive binding of differentially labeled cDNAs to the corresponding array elements. High resolution confocal fluorescence scanning of the array with two different wavelengths corresponding to the dyes used provides relative signal intensities and ratios of mRNA abundance for the genes represented on the array. **High-density oligonucleotide microarrays**: Array preparation: sequences of 16-20 short oligonucleotides (typically 25mer) are chosen from the mRNA reference sequence of each gene, often representing the unique part of the transcript. Light-directed, in situ oligonucleotide synthesis is used to generate high-density probe arrays containing over 300,000 individual elements. Target preparation: Total RNA from different tissues or cell populations is used to generate cDNA carrying a transcriptional start site for T7 DNA polymerase. During IVT, biotin-labeled nucleotides are incorporated into the synthesized cRNA molecules which is then fragmented. Each target sample is hybridized to a separate probe array and target binding is detected by staining with a fluorescent dye coupled to streptavidin. Signal intensities of probe array element sets on different arrays are used to calculate relative mRNA abundance for the genes represented on the array. Modified and reprinted with permission from Nature Cell Biology (Vol. 3, No. 8, pp. E190-E195) Copyright ©2001 Macmillan Publishers Limited.

Figure 2.1 The platforms of cDNA spotted arrays and oligonucleotide arrays (adopted from Wilson et al. [148]).

fluorescence intensities for each dye at each spot on the array. The ratio of red and green fluorescence intensities for each spot is expected to indicate the relative abundance of the corresponding molecule in the two target samples.

The oligonucleotide array technology has been commonly used to measure genome-scale expression levels. The technology was first developed by Fordor et al. [40]. Affymetrix GeneChip arrays further pioneered this technology and produces high density oligonucleotide based

DNA arrays [103]. The basic principles of manufacturing Affymetrix's GeneChips is that it uses photolithography and combinatorial chemistry to manufacture short single strands of DNA onto 5-inch square quartz chips. Unlike spotted cDNA arrays, the genes on the chip are designed based on sequence information alone. Each gene is represented by multiple short probes used to measure gene expression levels. Specifically, 11 to 20 perfect match (PM) and mismatch (MM) probe pairs are used to represent each gene, and PM-MM intensity differences are averaged for all probe pairs in a probe set to index expression level for each target gene.

In different microarray analyses, regardless of the platforms, typically either a one-color or two-color design will be used to measure mRNA abundance. A one-color design involves the hybridization of a single sample to each microarray after it has been labeled with a single fluorophore (for example, phycoerythrin, Cy3 or Cy5). In a two-color design, two samples (e.g., experimental and control) are labeled with different fluorophores (usually Cy3 and Cy5 dyes) and hybridized together on a single microarray. Although two-color designs have the potential to bring in bias and larger variations, techniques such as dye-reversed replicates (dye swaps or fluorophore reversals) can substantially improve the accuracy and sensitivity of gene expression measures. Compared to two-color designs, the advantage of one-color designs exists in their simplicity and flexibility. Furthermore, via biological and technical replicate assays, one-color designs can reduce data inconsistency across assays due to multiple sources of variability such as handling and processing [112].

2.1.2 Preprocessing and quality control of microarray data

The raw data from microarray analysis are image data. They need to be first converted to probe level numerical intensity data. This is then followed by background adjustment, normalization, and for some platforms, summarization. Background adjustment is to remove noise and non-specific hybridization. Normalization is to remove chip-to-chip variations possibly due to, e.g., chip design or color unbalance (for two-color designs) so measurements from different array hybridizations are comparable. For some platforms, e.g., Affymetrix GeneChips, a summarizing step is needed since gene expressions are represented by multiple probes and the summarizing step will combine multiple measures and produce a measure proportional to the amount of RNA transcript. Bioconductor has a complete set of functions that can be used for preprocessing

Affymetrix GeneChip data. Gentleman et al. [46] has a detailed discussion on these functions. Platforms by Agilent are also commonly used to produce genome-scale microarray data. In the following, we introduce the functions that are essential for preprocessing microarray data from Affymetrix GeneChips and Agilent data. For other platforms, readers can visit the bioconductor website, where several work flows are presented for different types of data, http://bioconductor.org/help/workflows.

Read in raw data. In Bioconductor, currently available functions cannot read in Affymetrix image data (i.e., the *DAT files*), but have functions to import numerical *CEL files*, which are intensity values produced by Affymetrix software. To read in *CEL files*, we need package `affy` and execute the following lines

```
library("affy")
Data = ReadAffy(celfile.path=)
```

The function `ReadAffy` will read in all *CEL files* in the specified working directory (specified by `celfile.path`) and creates an *AffyBatch* object. To read a particular set of *CEL files*, specify the file names as a character vector using `filenames=`.

To read in Agilent single color array data produced by the Agilent Feature Extraction image analysis software, we need the `limma` package with the following command lines

```
library("limma")
targets = readTargets("targets.txt")
x = read.maimages(targets, path="directory",
    source="genepix",green.only=TRUE)
```

The function `readTargets` will read in targets file, *targets.txt*, which contains information about the samples, and create a dataframe. Function `read.maimages` will read in image files contained in `"targets"` and create an `"RGList"` object `"x"`. In the above, `"path"` is the location of the targets file, and `"source"` is to provide the sources for the estimate of images. Here we assume that the images have been analyzed using GenePix to produce a .gpr file for each array, and that `targets.txt` has a column containing the names of the .gpr files. The argument `"green.only"` is to indicate that the images are from single color arrays.

Background adjustment. For data from Affymetrix GeneChips, two approaches are commonly used and available in Bioconductor to correct for background. One approach is to use PM intensities only due to the problems experienced when involving MM probes in the preprocessing steps [73]. A model to adjust PM intensities is designed based on the empirical distribution of probe intensities. The observed PM probe intensities are modeled as the sum of a normal noise component and a exponential signal component. It is noted that MM probe intensities are not corrected. The expectation of the exponential signal component (S) is given as

$$E(S|O = o) = a + b\frac{\phi(\frac{a}{b}) - \phi(\frac{o-a}{b})}{\Phi(\frac{a}{b}) + \Phi(\frac{o-a}{b}) - 1}$$
$$a = o - \mu - \sigma^2\alpha$$
$$b = \sigma,$$

where S is the signal following exponential distribution with mean α, O is the observed intensity, and μ and σ^2 are the mean and standard deviation of background noise. To avoid any possibility of negatives, truncated normal distribution at zero is utilized. To use this approach in Bioconductor to adjust background in PM probes,

```
> library("affy")
> data(Dilution)
> bgc.rma = bg.correct(Dilution, method="rma")
```

The data "Dilution" is an example data (as an AffyBatch object) and "rma" refers to the method noted above. Another approach for background adjustment uses both PM and MM probes, discussed in the Statistical Algorithms Description Document (2002) published by Affymetrix. In this approach, background values are first estimated and then each cell intensity is adjusted based on the estimated values. To estimate background values, the array is divided into 16 zones and for each probe cell, 16 distances to zone centers are calculated. For each zone, a background value is estimated as the average of lowest 2% intensities. Then the background value for a probe cell with coordinates (x, y), $b(x, y)$, is estimated as a weighted average of 16 zone background values,

$$b(x, y) = \frac{\sum_k w_k(x, y)b_k}{\sum_k w_k(x, y)}$$
$$w_k(x, y) = \frac{1}{d_k^2(x, y) + s_0}$$

$$d_k(x, y) = \text{distance to center of zone } k,$$

where b_k is the background value of zone k and s_0 is a smoothness parameter chosen to avoid non-informative weights. Cell intensities of PM and MM probes are then adjusted by the estimated background values with local noise taken into account. To perform the adjustment using the R function `bg.correct` in Bioconductor,

```
> bgc.mas = bg.correct(Dilution, method="mas")
```

Here, `"mas"` refers to the background adjustment method outlined above and included in the package Affymetrix Microarray Suite 5.0.

For data from Agilent, the package `Agi4x44PreProcess` is needed. To correct background noise, the function `backgroundCorrect` in the `limma` package can be used,

```
x.b = backgroundCorrect(x, method = , offset = )
```

An `RGList` object `"x.b"` will be produced, in which components R and G are background corrected. The object `"x"` is an `RGList` object noted earlier, which has intensities from different channels. The choice of `"offset"` will influence the determination on the number of differentially expressed genes, and thus a prior knowledge is needed to choose its value. In general, the larger the offset we choose, the larger the number of possibly differentiated genes. Different methods are available for background correction and users can also choose not to do background correction, `method="none"`. The default background correction method is to subtract the background intensity from the foreground intensity for each spot, `method="subtract"`. Among all the methods, `"normexp"` is suggested for the purpose of assessing differential expressions. This approach uses a convolution of normal and exponential distributions to fit to the foreground intensities with background intensities as a covariate, and then the estimated expected signal is the corrected intensity. This method along with a number of its variants [34, 107, 120, 128] result in a smooth monotonic transformation of the background subtracted intensities and all the corrected intensities are positive.

Normalization: The goal of normalization is to eliminate or reduce variations between arrays. A number of methods have been proposed and discussed in the literature for normalizing microarray data [10, 46]. The function `normalize` in bioconductor includes seven normalization methods. The methods built upon quantiles discussed in Worman et al. [149]

and Bolstad et al. [10] are utilized most often due to their simplicity and efficiency in computing and in general produce satisfactory results. We discuss these approaches in this section, in particular, the methods of quantile normalization and cubic-splines. For other methods, readers are referred to detailed discussions in Gentleman et al. [46].

The method of quantile normalizations discussed in Bolstad et al. [10] is based upon the concept of a quantile-quantile plot extended to n dimensions. If two arrays have the same distribution, then the quantiles plot of these two arrays should lie on a 45-degree diagonal line. This idea can be extended to n arrays. The corresponding algorithm is in the following several steps:

- step 1: after background adjustment, find the smallest intensity of each chip.

- step 2: average the values from step 1.

- step 3: replace each value in step 1 with the average computed in step 2.

- repeat steps 1 through 3 for the second smallest values, the third smallest values,..., and the largest values on each chip.

As seen in the description of the algorithm, the quantile normalization method is based on inverse transformation to obtain estimates of intensities, $x_i' = F^{-1}(G(x_i))$, where i indexes a specific location on an array, and G and F are estimated by the empirical distribution of each array and by the empirical distribution of the averaged sample quantiles, respectively.

To perform this normalization for Affymetrix GeneChip data using Bioconductor,

```
> norm.quant = normalize(Dilution, method="quantiles")
```

For data from Agilent,

```
x.norm.quant = normalizeBetweenArrays(x.b, method="quantile")
```

Object "x.b" is an RGList object and quantile normalization in this case is directly applied to the individual red and green intensities. Normalizing the A-values (the average of red and green intensities in the log2 scale) has been found to give more stable results:

```
x.b = backgroundCorrect(x, method = normexp, offset = 50)
MA.x.b = normalizeWithinArrays(x.b)
x.norm.quant = normalizeBetweenArrays(MA.x.b,
    method="Aquantile")
```

Function `normalizeWithinArrays` generates an object of class `MAList`.

The method of cubic-splines uses the quantiles from each array and the quantiles of geometric means (noted as the "target array") of each probes across all the arrays to fit a system of cubic splines to normalize the data [149]. All arrays are normalized to the target array. The target array does not have to be the geometric means and could also be a particular chip. This method is not available for Agilent data. The R codes to execute this normalization method for Affymetrix GeneChip data are

```
> norm.qspline = normalize(Dilution, method="qspline")
```

Summarization: This step is to produce expression level of each gene by combining multiple probe intensities for that gene. Different approaches are available to conduct the summarization. We focus on introducing two approaches available in the `affy` and `affyPLM` packages, the median polish approach and the one-step Tukey biweight-based method.

The median polish approach is to use an algorithm discussed in [111] to robustly fit a model describing intensity levels for a probe set. For a probe set k with $i = 1, \cdots, I_k$ probes and data from $j = 1, \cdots, J$ arrays, a multichip linear model is defined as,

$$\log_2\left(\text{Intensity}_{ij}^{(k)}\right) = \alpha_i^{(k)} + \beta_j^{(k)} + \epsilon_{ij}^{(k)},$$

where α_i is a probe effect and β_j is the estimated $\log_2$ expression value. The median polish approach is also available in the function `rma`.

The one-step Tukey bi-square weight algorithm calculates the signal value for a probe set at the $\log_2$ scale. Let m denote the median of signal values at I_k probes for probe set k and M be the median absolute deviation across those probes. The Tukey bi-square weight, $B(t)$, is calculated as

$$B(t) = \begin{cases} (1 - t^2)^2, & |t| < 1 \\ 0, & |t| \geq 1, \end{cases}$$

where $t = \frac{s-m}{5M+0.0001}$ with s denoting signal value. The signal value at probe set k is then calculated as the weighted average using the weight defined above, $S_k = \frac{\sum_{i=1}^{I_k} B(t_i)s_i}{\sum_{i=1}^{I_k} B(t_i)}$. Tukey bi-square weights have the property to determine a robust average unaffected by outliers.

In the `affy` package, these two approaches are incorporated into the function `expresso` as `medianpolish` and `mas`, respectively. The function `expresso` has different options for background correction, normalization and summary, for instance,

```
> eset = expresso(Dilution, normalize.method="qspline",
        bgcorrect.method="rma",pmcorrect.method="mas",
        summary.method="mas")
```

In the function `expresso`, `pmcorrect.method` refers to the methods for probe-specific background correction and `mas` means to adjust PM values by ideal mismatch values. The calculation of mismatch values is documented in [1]. Another method available in this function adjusts PM values by subtracting MM values. It also has an option of no adjustment, `pmonly`. For summarization, other methods are available besides the two approaches introduced above,

```
> express.summary.stat.methods()
[1] "avgdiff"  "liwong"   "mas"   "medianpolish" "playerout"
```

In the `affyPLM` package, the function `threestep` can be used to correct for background, normalize, and summarize the probes, e.g.,

```
> library("affyPLM")
> eset =  threestep(Dilution, background.method = "IdealMM",
        normalize="quantile", summary.method="tukey.biweight")
```

The background correction method, `"IdealMM"`, adjusts PM values by subtracting ideal mismatch values. Many other summary approaches are available in this function [9].

For data from Agilent, the function `avereps()` is used to average through replicate spots (or probes) and average across probes in a gene is not recommended.

2.2   DATA FROM NEXT GENERATION SEQUENCING

The RNA sequencing (RNA-Seq) technology makes it possible to discover and profile the transcriptome in any organism using deep-sequencing technologies. The transcriptome refers the complete set of transcripts in a cell along with their quantity. Understanding the transcriptome is essential for interpreting the functional elements of the genome, revealing the molecular compositions of cells and tissues, and understanding cell and tissue development and disease. We focus on bulk RNA-Seq data generation given its popular applications.

2.2.1 Data generation

Basically, RNA-Seq records the numerical frequency of sequences in a library population. In general, a population of RNA, such as poly(A)+, is converted to a library of cDNA fragments with adaptors attached to one or both ends. Each molecule is then sequenced in a high-throughput manner to obtain short sequences (i.e., reads) from one end (single-end sequencing) or both ends (paired-end sequencing). Each read is typically 30-400 base pairs, depending on the DNA-sequencing technology used. We will focus on RNA-Seq data generated from Illumina HiSeq 3000/4000. After sequencing, the resulting reads are either aligned to a reference genome or reference transcripts, or *de novo* assembled without sequenced genome to produce a genome-scale transcription map that consists of both the transcriptional structure and/or levels of expression for each gene [145].

Data generated from this technology has various advantages compared to microarray data. Unlike hybridization-based techniques, RNA-Seq is not limited to detecting transcripts that correspond to existing genomic sequence. Thus it is useful to identify novel genomic sequences for organisms of interest. RNA-Seq data have much lower background noise and are with high technical reproducibility compared to microarray data. RNA-Seq techniques are able to reveal locations of transcription boundaries at a single-base resolution. Furthermore, as noted in Wang et al. [145], short reads from RNA-Seq give information about how two exons are connected, whereas longer reads should reveal connectivity between multiple exons. These factors make RNA-Seq useful for studying complex transcriptomes.

2.2.2 Preprocessing and quality control of bulk RNA-Seq data

Raw RNA-Seq data usually come in a format of FastQ or SAM/BAM. Data in FastQ format are text files with each read composed of four lines and starting from @. The first line contains information on the sample, the second line is for the unaligned sequences (raw reads), the third line starts from '+' and in general no new information, and the last line records the quality of each read coded by ASCII characters.

The SAM (Sequence Alignment/Map) format is a standard format produced by alignment software. A SAM file is a tab-delimited text file that contains sequence of alignment data and a BAM file is its binary version. We can also have BAM data for unaligned reads. Another type of file in GTF (gene transfer format) is for the purpose of describing the structure of transcripts and how the transcripts are related to the corresponding genes.

Read in raw data. We focus on data in the FastQ format. To read in data, the `limma` package is needed. Then FastQ file names are read in using the R function `readTargets()`.

```
library(limma)
targets = readTargets()
```

The `targets` is a dataframe that has file names of FastQ data, which are then used in the subsequent steps to perform read alignment and summarization to assign reads to genes. A `targets` is in the following format,

	Sample	Group	InputFile	OutputFile
1	A1	A	A_1.fastq	A1.bam
2	A2	A	A_2.fastq	A2.bam
3	A3	A	A_3.fastq	A3.bam
4	B1	B	B_1.fastq	B1.bam
5	B2	B	B_2.fastq	B2.bam
6	B3	B	B_3.fastq	B3.bam

Examine the Quality of Data. To check the quality of RNA-Seq data, the ASCII characters recorded in the fourth line of each read in an FastQ file are used. The R function `qualityScores()` in a Bioconductor package Rsubread extracts the quality strings and convert them to quality scores ranged from -5 to 40. The higher the score, the higher the quality. Reads with low qualities should be excluded from further analyses.

```
library(Rsubread)
x = qualityScores(filename=targets$InputFile, offset=64,
        nreads=1000)
plot(colMeans(x))
```

In the above, `targets$InputFile` includes names of files that have sequence reads in FastQ format.

Reads alignment and counting. To perform read alignment, functions `buildindex()` and `align()` in the Rsubread package are needed. First we need to index the reference genome,

```
buildindex(basename="refIndex",reference="reference.fa")
```

The file `"reference.fa"` contains all the reference sequences in an FastA format, and the created index file will be saved in `"refIndex"`. Next, we align FastQ files with the reference sequence:

```
align(index="refIndex",readfile1=targets$InputFile,
+ output_file="aligned.bam")
```

The function `align` can take FastQ and FastA read files. The names of the read files are given by `target$InputFile`. The aligned reads are saved in `"aligned.bam"`. After the reads aligned, we are now ready to assign reads to genomic features using the function `featureCounts`.

```
readCounts = featureCounts(files="aligned.bam",
+ annot.ext="annotation.gtf")
```

The mapped BAM data and the annotation data in the file `annotation.gtf` are the input, and the output is an R list object which includes the counts to each genomic feature and their annotations.

Note that up to this point, the data are counts for each genetic feature, e.g., a gene. Due to this, in subsequent analyses, different types of methods have been proposed with one type directly implementing the counts and the other transforming the counts and analyzing at the continuous scale. There exist a wide range of approaches currently available for analyzing RNA-Seq data. Among these methods, it seemed that two methods performed better. One approach is variance stabilization transformation implemented in the functions `voom` and `vst` [131] followed by linear regressions included in the function `lmFit` in the package `limma`. The other is a non-parametric method built upon Wilcoxon rank statistic included in the Bioconductor package SAMseq [96].

## CHAPTER 3

Genome-scale epigenetic data

DNA methylation is a modification of DNA by a covalent addition of a methyl group to the DNA at the 5-position of cytosine followed by a guanine. DNA methylation can be inherited. Histone modifications are chemical modifications due to the less structured N-terminal domains of all core histones protrudent from the core histone [87]. Both epigenetic changes have the potential to regulate gene functions. In this book, we focus on DNA methylation.

3.1 DATA GENERATION

The commonly used DNA methylation measurement platforms are from Illumina (Illumina, Inc., San Diego, CA, USA), including the Infinium HumanMethylation27 BeadChip, HumanMethylation450 BeadChip, and MethylationEPIC, among which Infinium HumanMethylation27 Bead-Chip has been withdrawn from the market, and MethylationEPIC appeared on the market in the year of 2016. A HumanMethylation450 BeadChip (also called 450K BeadChip/array) uses two different probe designs, Infinium Types I and II probes, to analyze methylation status on 485,577 methylation sites including cytosine-phosphate-guanine (CpG) sites across the human genome. The purpose of using two probes is to enhance the depth of coverage for methylation analyses. The Type I probes are the same as those employed in the HumanMethylation27 BeadChip. This type of probes uses two 50bp probes with one for methylated (M) sites, and the other for unmethylated (U) sites. The Type II probes are new to the 450K array. It uses one probe per locus and different dye colors (green and red) to differentiate between M and U signals. MethylationEPIC uses the same technique as in HumanMethylation450 BeadChip, but covers a much larger number of methylation sites compared to the 450K array. DNA methylation of more than 850,000 sites is measured by use of a MethylationEPIC array.

Two formats of DNA methylation data are available. One format is IDAT (Intensity DAT) and the other is a text document. The IDAT format is used to store BeadArray data from profiling platforms. Data in this format are direct output from the scanner and are summarized intensities for each probe-type on an array in a compact manner. Data in the text format represent DNA methylation levels, denoted as β values. A methylation level is calculated as the proportions of intensity of methylated (M) over the sum of methylated and unmethylated (U) sites $(\beta = M/(c + M + U])$, where parameter c is a small number introduced for the situation of too small $M + U$ to avoid underflow.

3.2 QUALITY CONTROL AND PREPROCESSING OF DNA METHYLATION DATA

Different methods have been proposed for DNA methylation data quality control and preprocessing. We focus on a pipeline proposed by Lehne et al. [95] and the method included in an R package `minfi` [42]. Both approaches can be applied to DNA methylation data from HumanMethylation450 BeadChips as well as MethylationEPIC.

3.2.1 The control probe adjustment and reduction of global correlation pipeline (CPACOR)

This is a quality control and preprocessing pipeline and it utilizes information on control probes to adjust for technical bias. The diagram of the pipeline is in Lehne et al. [95] and included in Figure 3.1. Multiple tasks are carried out in this pipeline and components related to DNA methylation data quality control and preprocessing include the step of initial quality control and steps marked by solid arrows in Figure 3.1. A set of R programs following this pipeline are available in [95] and can be downloaded from the publisher's online source, `https://static-content.springer.com/`. In this section, instead of going through the whole set of programs in CPACOR, we discuss key R functions in the pipeline related to the main topic of this section. In particular, the R functions utilized in `1_intensities.sh` and `2_betas.sh` in the program set. A data

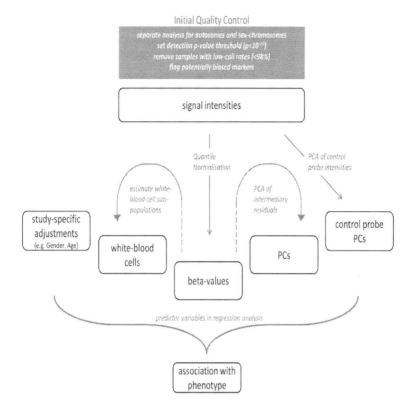

Figure 3.1 The pipeline to pre-process DNA methylation data [95].

set available in the Gene Expression Omnibus (GEO) database, https://www.ncbi.nlm.nih.gov/geo/query/acc.cgi?acc=GSE69636, will be used to demonstrate the key functions. Since some R packages in these two programs have been updated and for the purpose of illustration, we revised the two programs accordingly with an application to the GEO data set and included them at the end of this chapter for users' convenience. For other parts in the pipeline, we give a brief discussion later in this section.

This pipeline uses raw IDAT file and can be called in using R function read.metharray.exp in the minfi package. Function bgcorrect.illumina is used to perform background subtraction:

```
library(minfi)
rgSet = read.metharray.exp("idat/", verbose=TRUE)
RGset = bgcorrect.illumina(rgSet)
```

The object `rgSet` is an object of class representing unprocessed data from Illumina methylation arrays. It includes all two-color (red and green) IDAT files in the directory given by `"idat"`. The CPACOR pipeline performs initial quality control for the purpose of removing single nucleotide polymorphism (the methylation sites are labeled as "rs"), background subtractions, and marking low quality intensity values as "NA". A low quality intensity is determined if detection P values $\geq 10^{-16}$.

Quantile normalization is usually applied to reduce technical biases. After comparing a number of approaches, Lehne et al. [95] recommend conducting quantile normalization on intensity values instead of β values and such a normalization procedure is applied to DNA methylation measures in each of six categories determined by probe types, color channel, and Methylated/Unmethylated subtypes: Type-I M red, Type-I U red, Type-I M green, Type-I U green, Type-II red, and Type-II green. For instance, "Type-I M red" represents intensity measures of methylated type I probes on the red channel. The information on probe types is available from the Illumina manifestation file and can be downloaded from Illumina's website. Methylated and unmethylated signal intensity values for each channel are available in IDAT files and can be extracted as numeric matrices using the R functions `getGreen` and `getRed`. Before performing quantile normalization, these intensity values should have been initially quality controlled. The R function for quantile normalization for each of the six categories is

```
Intensity = normalizeQuantiles(data)
```

Since the above normalization function is applied to each of the six categories, function argument `"data"` refers to a numeric matrix of methylated or unmethylated intensity values corresponding to a specific probe type and channel, e.g., `TypeII.Green` as in the program included at the end of this Chapter. After normalizing the intensity data, β values are then calculated using the intensities, following $\beta = M/(c + M + U])$, where c is usually taken as 100. Detailed codes of intensity normalizations are in the program `2_betasRev.R`

Compared with other existing approaches for preprocessing, the significant difference in this CPACOR pipeline is its utilization of control probes. Control probes are included in the Illumina methylation arrays to assess multiple aspects of the chemistry such as bisulfite conversion contributing to measurements of DNA methylation. Involving control probes in preprocessing is to further reduce technical biases that cannot be removed by quantile normalization. Given the high correlations among the control probes, a principal component analysis on control probes is performed to identify latent factors potentially contributing to the variation in intensity measures after initial quality controls.

```
PCA = prcomp(na.omit(ctrl.all))
ctrlScores = PCA$x
```

The function `prcomp()` infers the principal components in methylation intensities of control probes. Lehne et al. [95] suggest 30 components be extracted and these components are used in subsequent analyses to adjust for technical bias. For instance, regressing quantile normalized and logit transformed DNA methylation β values (i.e., the M values) on the principal components to adjust for technical biases, and then use the regression residuals in subsequent analyses. M values are suggested for the purpose of reducing heteroscedasticity by Du et al. [28]. On the other hand, if DNA methylation in the analyses is treated as "dependent" variables, then the principal components can be included as "independent" variables in the analytical models, e.g., linear regressions. Based on our experience and taking into account possible small sample sizes that can happen in many studies, 15 components in general are sufficient to capture additional technical biases.

Furthermore, as noted earlier, in the pipeline proposed by Lehne et al. [95], additional steps are discussed, which include creating additional principal components using residuals after adjusting for cell compositions as well as factors specific to studies such as gender and age. These factors are in general included in subsequent analyses. In addition, the residuals in this case include unknown factor effects, and thus including the inferred principal components in subsequent analyses can potentially cause concerns of over-fitting. We thus suggest skip the subsequent steps in the pipeline. In summary, this pipeline includes the following three steps related to quality control and preprocessing: 1) initial quality control, 2) quantile normalization, and 3) principal component analysis on control probes.

3.2.2 Quantile normalization with ComBat

Based on β values calculated from quantile normalized intensities as done in the CPACOR pipeline, this approach, quantile normalization with R function ComBat, applies R function `ComBat` to the base 2 logit transformed β values to further diminish heterogeneity caused by, for instance, different batches of DNA methylation data. The method implemented in the `ComBat` is an empirical Bayes approach that incorporates systematic errors assuming their similar impact on DNA methylation across different batches. The batch effects are estimated by pooling information across CpG sites in each batch, which are then used to adjust batch effects in DNA methylation at each CpG site. In the following, we present the R codes implementing `ComBat`:

```
Mvalues = log((beta/(1-beta)),2)
mod = model.matrix(~1, data=pheno)
combatM = ComBat(data=Mvalues, batch=batch, mod=mod,
    par.prior=TRUE, prior.plots=TRUE)
```

The first line calculates M values, which are approximated by logit transformed β values at base 2. The `ComBat` function requires a model matrix, `model.matrix`, to include factors (in the data `pheno`) potentially contributing to technical variations in DNA methylation, in addition to batches. In the above, only intercept defined by ~ 1 is in the model matrix. In the function `ComBat`, `batch` is a vector indicating batches of DNA methylation data and used as a covariate.

After preprocessing the data with either CPACOR pipeline or the `combat` approach, it is suggested that CpG sites with nearby probe SNPs are excluded from subsequent analyses. This is under the consideration that DNA methylation at those CpG sites are potentially contaminated and thus do not correctly reflect DNA methylation levels.

3.3   CELL TYPE COMPOSITION INFERENCES

In many situations, genome-scale genetic and epigenetic assessments used in association or prediction studies are based on whole blood samples. However, due to cellular heterogeneity in whole blood, the inferred associations or predictions could be contaminated by biased assessment on gene expressions or DNA methylation caused by such heterogeneity. Removing or at least reducing the bias to the maximum extent is desired. In large epidemiological studies, it is not feasible to isolate and profile every individual cell subset for cell counts using technologies such as flow cytometry. Thus, several algorithms have been developed to estimate and adjust for cellular heterogeneity in whole blood. Recently developed methods for cell type composition estimations mainly utilize genome-scale DNA methylation. In this section, we briefly review these developed approaches used to infer cell compositions or to adjust the effects of cell type heterogeneity in association or prediction analyses. A comprehensive review of a set of methods along with corresponding R programs on cell type compositions assessment can be found in Kaushal et al. [85].

3.3.1 Reference-based methods

This type of methods require a reference database as a gold standard. Houseman et al. [71] developed a method for estimating cell type proportions based on DNA methylation data measured at the genome scale. This approach capitalizes on the idea that differentially methylated regions (DMRs) can serve as a signature (gold standard) for the distribution of different types of white blood cells, since the measured DNA methylation is for a specific cell that is known. It uses these DMRs as a surrogate in a regression-calibration-based technique to estimate cell distributions in a mixture of cells. Regression calibrations can lead to bias estimates, thus an external validation data is used to calibrate the model and to correct for the bias, using measurement error modeling. Their method was specifically developed for the Infinium HumanMethylation27 BeadChip array (Illumina, Inc., San Diego, CA, USA).

The basic modeling in Houseman et al. includes two parts. The first part describes associations between DNA methylation at a set of CpG sites and covariates of interests via linear regressions, and the second part links two sets of coefficient parameters in the linear regressions. Following the notation in Houseman et al., for the first part,

$$
\begin{aligned}
\boldsymbol{Y}_{0h} &= \boldsymbol{B}_0 \boldsymbol{w}_{0h} + e_{0h} \\
\boldsymbol{Y}_{1i} &= \boldsymbol{B}_1 \boldsymbol{z}_{1i} + e_{1i},
\end{aligned}
\tag{3.1}
$$

where $\boldsymbol{Y}_{0h}$ is an $m \times 1$ vector composed of DNA methylation on a set of pre-selected CpG sites measured from a purified blood sample consisting of a single cellular population, for instance, monocytes or B cells, and vector $\boldsymbol{w}_{0h}$ is a $d_0 \times 1$ vector with the first component being one and the remaining composed of 1's and 0's indicating cell type of specimen h. Thus the first equation describes the expected cell composition of specimen h, represented by $\boldsymbol{B}_0$ with dimension $m \times d_0$, on each CpG site for its corresponding cell type. In the second equation, $\boldsymbol{Y}_{1i}$, like $\boldsymbol{Y}_{0h}$, is an $m \times 1$ vector composed of DNA methylation on the same set of pre-selected m CpG sites. The difference is that in this case DNA methylation is measured from a whole blood sample of subject i, $i, = 1, \cdots, I$, which is in general contaminated by various cell types. Vector z_{1i} is with dimension d_1 with the first element being 1 and the remaining for phenotypes, exposures, or health outcomes of the ith subject. For instance, if z_{1i} represents the status of a disease (Yes/No), then z_{1i} is composed of $(1, \delta)$ with dimension $d_1 = 2$ and $\delta = 0$ or 1. Parameter $\boldsymbol{B}_1$ evaluates the association of $\boldsymbol{Y}_{1i}$ with z_{1i} with the first column being the intercepts.

Next, we connect observed data, $\boldsymbol{Y}_{1i}$, with the gold standard, Y_{0h}, as the following,

$$\boldsymbol{B}_1 = 1_m \boldsymbol{\gamma}_0^T + \boldsymbol{B}_0 \boldsymbol{\Gamma} + \boldsymbol{U}, \qquad (3.2)$$

where 1_m is a vector of 1's with size m, and $\boldsymbol{\gamma}_0$ is a d_1-dimensional vector representing the intercepts. Parameter $\boldsymbol{\Gamma}$ is a matrix with dimension $d_0 \times d_1$ assessing the association of $\boldsymbol{B}_1$ with $\boldsymbol{B}_0$, which measures to what extent the association of $\boldsymbol{Y}_{1i}$ with z_{1i} was due to cell heterogeneity.

To estimate cell type compositions for one subject, subject i, we revise the second equation in (3.1) by only keeping the intercepts, i.e.,

$$\boldsymbol{Y}_i^* = \tilde{\boldsymbol{B}}_1 + e_{1i}^*.$$

In this case, $\tilde{\boldsymbol{B}}_1$ is a vector of size m. Consequently, (3.2) becomes

$$\tilde{B}_1 = 1_m [\boldsymbol{\gamma}_0^*]^T + \tilde{\boldsymbol{B}}_0 \boldsymbol{\Gamma}^* + \boldsymbol{U}^*, \qquad (3.3)$$

with vector $\boldsymbol{\Gamma}^*$ of size d_0 representing cell type compositions of d_0 cells for subject i. Denoting by $\tilde{\boldsymbol{B}}_0 = (1_m, \beta_0)$ and $\tilde{\boldsymbol{\Gamma}}^* = (\boldsymbol{\gamma}_0^*, \boldsymbol{\Gamma}^*)$, we estimate $\tilde{\boldsymbol{\Gamma}}^*$ by minimizing squared errors $|\boldsymbol{Y}^* - \tilde{\boldsymbol{B}}_0 \tilde{\boldsymbol{\Gamma}}^*|^2$ with respect to $\tilde{\boldsymbol{\Gamma}}^*$. Since DNA methylation assessment typically focuses on comparison of methylated to unmethylated CpG dinucleotides, not quantifying actual amounts of DNA, the vector $\boldsymbol{\Gamma}^*$ represents cell mixtures with respect to distributions, and in particular, the estimated $\boldsymbol{\Gamma}^*$ is proportional to the compositions of d_0 cells.

Jaffe and Irizarry [79] further extended the method in Houseman et al. [71] and tailored it to fit DNA methylation data from the Illumina HumanMethylation450 BeadChip (450K) array as well as the Human-Methylation27 BeadChip (27K) array. The algorithm in Houseman et al. proposed 500 CpG sites to estimate cell mixture proportions from the Illumina 27k array. The modification of Jaffe and Irizarry was necessary because of the existence of probe SNPs in the suggested 500 CpG sites and the inconsistency of CpG sites between the 27K and 450K arrays. In addition, the flow-sorted data of the six adult male subjects were used as references when DNA methylation was measured in peripheral blood [79].

The method by Jaffe and Irizarry has been included in an R package, `minfi`. Kaushal et al. [85] has a detailed example to demonstrate this approach, using a data set available at https://www.microsoft. com/en-us/download/details.aspx?id=52345. Here we do not repeat the whole example program and focus on an explanation of some key functions and their connection with the method discussed earlier in this section.

```
lib = c("minfi","quadprog","FlowSorted.Blood.450k",
    "IlluminaHumanMethylation450kmanifest",
    "IlluminaHumanMethylation450kanno.ilmn12.hg19")
lapply(lib, require, character.only = TRUE)
grSet = read.table("input_data.txt",header=T)
grSet = data.matrix(grSet)
referenceMset = get('FlowSorted.Blood.450k.compTable')
cell = c("CD8T","CD4T", "NK","Bcell","Mono","Gran","Eos")
compData = minfi:::pickCompProbes(referenceMset,
            cellTypes=cell)
coefs = compData$coefEsts
coefs = coefs[ intersect(rownames(grSet), rownames(coefs)), ]
rm(referenceMset);
cellProp = minfi:::projectCellType(grSet[rownames(coefs), ],
            coefs)
```

The first four lines are to specify the libararies for packages and reference database that go with `minfi`, and `grSet` is the cell-type "contaminated" DNA methylation data set with n rows (number of CpG sites) and I (number of subjects) columns. The five lines after `grSet = data.matrix(grSet)` is to select the reference database, which is 'FlowSorted.Blood.450k.compTable', specify the cell types for composition estimation, select CpGs from the reference database to estimate coefficients $\tilde{B}_0$, and extract coefficients $\tilde{B}_0$ for the selected CpG sites. The function `projectCellType` is to obtain an estimate of cell type proportions, `cellProp`.

For DNA methylation in newborns, adulthood references cannot be used due to substantial differences in cell compositions between blood in newborns and that in adults [4]. Bakulski et al. [4] proposed reference data for newborns based on cord blood, which is available in the `minfi` package. In addition to cord blood, several drops of capillary blood on pre-printed cards (Guthrie cards) are often collected from newborns by puncturing the heel of the newborn babies. Collection of Guthrie cards is easier in comparison of cord blood collection and a routine process in many places. Furthermore, storage of Guthrie cards are easier than cord blood. Thus, DNA methylation measured from Guthrie cards is increasingly utilized in epigenetic studies. However, in cell composition assessment, a recent study indicated a potential bias of applying references data designed for cord blood to Guthrie cards [80]. Between cord blood and Guthrie cards, less than 10% CpGs showed a correlation of 0.5 or higher in DNA methylation, implying potential disagreement in cell compositions between these two sources. Thus cautions are needed when inferring cell heterogeneity using DNA methylation measured in Guthrie cards but basing on cord blood references.

Other approaches for cell mixture inferences are also available that utilize reference databases. The method of removing unwanted variation (RUV) is one of such methods [106]. Although this approach needs a reference database, it does not estimate cell type proportions. Instead, this approach utilizes negative control probes and performs principal component (PC) analyses on these probes followed by orthogonal rotations on the PCs to identify factors due to unmeasured confounders. These factors are then included in subsequent analyses to adjust for cell type effects. The negative control probes were chosen as the top 500 CpG sites from the reference database of DNA methylation known to be correlated with the cell types [106, 85]. To implement this method in R, two libraries are needed, library `psych` for principal component analyses and rotations (`prcomp`), and library `pracma` has various functions which can be used to calculate scores based on loadings inferred from control probes (`pinv`). The complete program can be downloaded from GitHub, https://github.com/akhileshkaushal/Cell-Proportions/tree/master with an application to a DNA methylation data set available in Houseman et al. [71].

```
library(data.table)
library(psych)
library(pracma)
betat = t(beta)
pca_mval = prcomp(betat,center=T,scale.=T)
plot(pca_mval, type = "l")
#Based upon scree plot we can choose number of PC's##
```

```
rawLoadings = pca_mval$rotation[,1:7] %*% diag(pca_mval$sdev,
            7, 7)
rotatedLoadings = varimax(rawLoadings,normalize = TRUE,
            eps = 1e-5)$loadings
invLoadings = t(pracma::pinv(rotatedLoadings))
scores = scale(beta1_t) %*% invLoadings
```

In the codes above, `betat` represents a matrix with entries of DNA methylation in β values such that each row is for one CpG site and each column is for one subject. Varimax orthogonal rotation (`varimax`) is used to maximize the projection loadings. The scores, calculated as the standardized DNA methylation data multiplied by the transpose of generalized inverse of the rotated loading matrix (`t(pracma::pinv(rotatedLoadings))`), have information on cell compositions for each subject to be used in subsequent analyses.

3.3.2 Reference-free methods

Methods on cell type heterogeneity that do not rely on any reference databases have also been developed. An advantage of these non-reference-based methods is that they can be applied to other tissues in addition to blood. Zou et al. [167] developed a non-reference-based method, FaST-LMM-EWASher. This approach is built upon linear mixed models with top principal components as the covariates. Another set of methods infer latent variables for cell type compositions, which are then included in association assessments. These methods include RefFreeEWAS [72] and its improved version (RefFreeCellMix, [70]), surrogate variable analysis (SVA, [93]), and ReFACTor [117]. Among all these methods, from findings in recent studies [85], we focus on three approaches, RefFreeCellMix, SVA, and ReFACTor. These three approaches overall perform well when no underlying confounding effects exist. However, when latent variables are correlated with observable variables, SVA outperforms the other two approaches, RefFreeCellMix and ReFACTor [85]. Programs of these three methods with examples are available in Kaushal et al. [85], and our focus here is to explain some key functions.

The methods in `RefFreeEWAS` and `RefFreeCellMix` both utilize singular value decompositions (SVDs) and extract latent subject and cell-specific effects. `RefFreeCellMix` is an improved version of `RefFreeEWAS`. `RefFreeEWAS` applies singular value decomposition (SVD) to decompose the residuals of unadjusted linear models along with unadjusted linear coefficient estimates, and estimates latent subject and cell-specific effects. Bootstrap estimates for coefficient standard errors are used to account for the correlation in the error structure. `RefFreeCellMix`, on the other hand, directly works on DNA methylation data to avoid reliance on phenotypic data and possible over adjustment on cell type compositions. It uses a variant of non-negative matrix factorization to decompose the total methylation sites into CpG-specific methylation states for a pre-specified number of cell types and subject-specific cell-type distributions [70],

$$y = M\Omega^T,$$

where Y is an $m \times n$ matrix representing DNA methylation data in β values, M is an unknown $m \times K$ matrix, and Ω is an $n \times K$ matrix. Matrices M and Ω are constrained to lie in the unit interval and the values in each row of Ω represent presumed cell type proportions summed to at most 1. The purpose of specifying the number of cell types is to reduce computational burden. To estimate Ω using `RefFreeCellMix`, the library `RefFreeEWAS` is needed.

```
library(RefFreeEWAS)
cell = RefFreeCellMix(y,K=7,iters=5)
```

Argument y is an $m \times n$ matrix representing DNA methylation data in β values at m CpG sites in n subjects, and `iters` specifies the number of iterations of the estimating process, and K specifies the assumed number of cell types. It is possible that in some cases the number of cells of interest is unknown. To estimate K, Houseman et al. [70] proposed a method built upon bootstrap samples to minimize mean bootstrap deviance (with upper and lower quartiles trimmed). Given the high dimensionality, this approach can be rather slow. Two R functions `RefFreeCellMixArray` and `RefFreeCellMixArrayDevianceBoots` are implemented to estimate K,

```
yshortTest = y[1:500,]
testArray = RefFreeCellMixArray(yshortTest,Klist=1:5,iters=5)
testBootDevs = RefFreeCellMixArrayDevianceBoots(testArray,
               yshortTest)
which.min(apply(testBootDevs[-1,],2,mean,trim=0.25))
```

The number of cell types K is estimated based on a subset of CpG sites. In the above, DNA methylation at 500 CpG sites are used. The argument Klist specifies a vector of values of K, for each value of K a deviance is calculated, and `iters` specifies the number of iterations needed to calculate deviance. Finally, the trimmed mean deviance across bootstrap samples is calculated by `apply(testBootDevs[-1,],2,mean,trim=0.25)`, and the minimum is identified by use of the `which.min` function.

Surrogate variable analysis (SVA), based on SVDs of residuals in linear regressions, uses permutations to identify eigen-vectors and consequently infer potential confounding factors (surrogate variables). The residuals are obtained after regressing DNA methylation on phenotypic variables of interest. SVA was initially applied to gene expression data, and utilized the concept of expression heterogeneity while estimating surrogate variables. Expression heterogeneity (EH) refers to certain plausible biological profiles of the subject, which may not be captured by the observed variables. SVA decomposes the residual matrix and utilizes permutations to identify statistically significant eigen-vectors, representing expression heterogeneity. These eigen-vectors are called eigengenes, and surrogate variables are then inferred based on theses "eigengenes". Surrogate variables estimated using SVA have the potential to cover information on cell types in DNA methylation from blood cells. A Bioconductor package is available to estimate surrogate variables using this approach [94]. To infer surrogate variables,

```
mod1 = model.matrix(~x1)
mod0 = model.matrix(~1)
sv = sva(edata,mod1,mod0,n.sv=NULL,method="two-step")
```

Surrogate variables are estimated based on logit transformed β values of DNA methylation, `edata`, a matrix with the variables in rows and samples in columns. A model used to fit the data is `mod1`, and the null model is `mod0`. The argument `method` specifies the mehtod used to estimate the surrogate variables. In the above, the `two-step` approach is chosen. It is based on a subset of rows affected by unmodeled dependence [92]. The other two available approaches are `"irw"` (the iteratively re-weighted approach) and `"supervised"` (the supervised approach). The iteratively re-weighted method is for empirical estimation of unknown factors, and the supervised method is for the situation that control probes are known (i.e., probes such that genes are unlikely to be differential expressed [92]). Finally, `n.sv` sets the number of surrogate variables. Here, `n.sv` is `NULL` and in this case the number of surrogate variables will be estimated based on the data. The inferred surrogate variables can be extracted by `sv$sv` and applied in subsequent analyses as covariates used to adjust for confounding effects from cell types along with other unknown factors. Kaushal et al. [85] demonstrated that using SVA to adjust for cell composition heterogeneity potentially leads to the highest power with respect to statistical testing on variables of interest.

Finally, the ReFACTor approach is based on principal component analyses on a set of potentially informative CpG sites [117]. ReFACTor implements a variant form of principal component analysis (PCA) to adjust for the cell type effects. This method assumes that a small number of methylation sites are affected by underlying cell mixtures, and excludes CpGs if variations in DNA methylation are not large enough (the default cutoff is a standard deviation of 0.02 in β values). In Kaushal et al. [85], to avoid excessive exclusions, CpGs such that their standard deviations in the lower 5th percentile were excluded. This method implemented in the R package by default searches for top 500 most informative methylation sites and performs PCA with a fixed number of components on these CpG sites to obtain the components. These ReFACTor components can be used as covariates to adjust for the effects due to cell type heterogeneity.

The implementation of the ReFACTor method in R includes the following three parts. The first part is to obtain DNA methylation not explained by known factors.

```
SD = apply(y, 1, sd, na.rm = F)
Include = which(SD>=quantile(SD,0.05))
Mark = y[include,]
names = rownames(Mark)
for (site in 1:nrow(Mark))
{
  model = lm(Mark[site,] ~ z)
  Mark[site,] = residuals(model)
}
```

In the above, standard deviations of each CpG sites are calculated by the apply function, CpGs with small standard deviations are excluded, and residuals of DNA methylation at the retained CpGs not explained by covariate z are calculated based on linear regressions.

The second part is to identify CpGs potentially influenced by cell compositions (the R codes are below). The approach first estimates the principal components based on the residuals, and then using the loadings (coeff) and scores (score) calculates the k-th rank approximation (Bn) of the centered Mark (An). The distance between An and Bn is determined by an Euclidean distance on standardized An and Bn. The top 500 CpGs with the lowest distance are selected, as they are likely the most representative CpGs reflecting An. In this process, k is the presumed number of cell types in the data, and it is chosen as 7, which has been the commonly adopted number of cell types.

```
k = 7
pcs = prcomp(scale(t(Mark)))
coeff = pcs$rotation
score = pcs$x
x = score[,1:k]%*%t(coeff[,1:k])
An = scale(t(Mark),center=T,scale=F)
Bn = scale(x,center=T,scale=F)
An = t(t(An)*(1/sqrt(apply(An^2,2,sum))))
Bn = t(t(Bn)*(1/sqrt(apply(Bn^2,2,sum))))
distances = apply((An-Bn)^2,2,sum)^0.5
dsort = sort(distances,index.return=T)
ranked_list = dsort$ix
sites = ranked_list[1:500]
```

The last step is to obtain a set of scores correlated with the real cell type proportions. This is achieved by performing principal component analyses on the residuals of the 500 selected CpGs, and the first k columns of the scores, first_score, are the proposed set of scores representing cell type compositions to be included in subsequent analyses.

```
pcs = prcomp(scale(t(Mark[sites,])))
score = pcs$x
first_score <- score[,1:k]
```

3.4 APPENDIX – MODIFIED PROGRAMS IN THE CPACOR WITH AN APPLICATION

The following programs are the programs in Lehne et al. [95] with some modifications and an application to a GEO IDAT data set at https://www.ncbi.nlm.nih.gov/geo/query/acc.cgi? acc=GSE69636. The following program can also be downloaded from the author's website, https://www.memphis.edu/sph/contact/faculty_profiles/zhang.php.

```
# 1\_IntensitiesRev.R
# Retrieves signal-intensities from Illumina idat-files. Returns
# detection p-values and signal intensities for genomic CpGs
# and control probes. Originally written by Benjamin Lehne
# (Imperial College London) and Alexander Drong (Oxford University).
# Modified by Hongmei Zhang (University of Memphis) with an
# application  to a GEO dataset.
```

```
require(minfi)
require(IlluminaHumanMethylation450kmanifest)

#read data
rgSet = read.metharray.exp("idat/",verbose=TRUE)
RGset = bgcorrect.illumina(rgSet)# Illumina background subtraction

# Type II probes
TypeII.Name = getProbeInfo(RGset, type = "II")$Name
TypeII.Green = getGreen(RGset)[getProbeInfo(RGset, type = "II")
$Address,]
TypeII.Red = getRed(RGset)[getProbeInfo(RGset, type = "II")
$Address,]
colnames(TypeII.Red) = sampleNames(RGset)
colnames(TypeII.Green) = sampleNames(RGset)

# Type I probes, split into green and red channels
TypeI.Green.Name = getProbeInfo(RGset, type = "I-Green")$Name
TypeI.Green.M = getGreen(RGset)[getProbeInfo(RGset, type =
"I-Green")$AddressB,]
rownames(TypeI.Green.M) = TypeI.Green.Name
colnames(TypeI.Green.M) = sampleNames(RGset)
TypeI.Green.U = getGreen(RGset)[getProbeInfo(RGset,
type = "I-Green")$AddressA,]
rownames(TypeI.Green.U) = TypeI.Green.Name
colnames(TypeI.Green.U) = sampleNames(RGset)

TypeI.Red.Name = getProbeInfo(RGset, type = "I-Red")$Name
TypeI.Red.M = getRed(RGset)[getProbeInfo(RGset, type = "I-Red")
$AddressB,]
rownames(TypeI.Red.M) = TypeI.Red.Name
colnames(TypeI.Red.M) = sampleNames(RGset)
TypeI.Red.U = getRed(RGset)[getProbeInfo(RGset, type = "I-Red")
$AddressA,]
rownames(TypeI.Red.U) = TypeI.Red.Name
colnames(TypeI.Red.U) = sampleNames(RGset)

#BSC1 control probes
BSCI.Green.Name = getProbeInfo(RGset, type = "Control")[16:18,]
$ExtendedType
BSCI.Green = matrix(NA_real_, ncol = ncol(RGset), nrow =
length(BSCI.Green.Name), dimnames = list(BSCI.Green.Name,
sampleNames(RGset)))
BSCI.Green[BSCI.Green.Name,] = getGreen(RGset)[getProbeInfo(RGset,
type = "Control")[16:18,]$Address,]
BSCI.Red.Name = getProbeInfo(RGset, type = "Control")[22:24,]
$ExtendedType
BSCI.Red = matrix(NA_real_, ncol = ncol(RGset),
nrow = length(BSCI.Red.Name),
dimnames = list(BSCI.Red.Name, sampleNames(RGset)))
BSCI.Red[BSCI.Red.Name,] = getRed(RGset)[getProbeInfo(RGset, type =
```

```
"Control")[22:24,]$Address,]


#BSC2 control probes
BSCII.Red.Name = getProbeInfo(RGset, type = "Control")[28:31,]
$ExtendedType
BSCII.Red = matrix(NA_real_, ncol = ncol(RGset),
nrow = length(BSCII.Red.Name),
dimnames = list(BSCII.Red.Name, sampleNames(RGset)))
BSCII.Red[BSCII.Red.Name,] = getRed(RGset)[getProbeInfo(RGset,
type = "Control")[28:31,]$Address,]


#STAINING
stain.Red.Name =getProbeInfo(RGset, type = "Control")[2,]
$ExtendedType
stain.Red = matrix(NA_real_, ncol = ncol(RGset),
nrow = length(stain.Red.Name),
dimnames = list(stain.Red.Name, sampleNames(RGset)))
stain.Red[stain.Red.Name,] = getRed(RGset)[getProbeInfo(RGset,
type = "Control")[2,]$Address,]
stain.Green.Name = getProbeInfo(RGset, type = "Control")[4,]
$ExtendedType
stain.Green = matrix(NA_real_, ncol = ncol(RGset),
nrow = length(stain.Green.Name),
dimnames = list(stain.Green.Name, sampleNames(RGset)))
stain.Green[stain.Green.Name,] = getGreen(RGset)[getProbeInfo(RGset,
type = "Control")[4,]$Address,]


#EXTENSION
extensionA.Red.Name =getProbeInfo(RGset, type = "Control")[7,]
$ExtendedType
extensionA.Red = matrix(NA_real_, ncol = ncol(RGset),
nrow = length(extensionA.Red.Name),
dimnames = list(extensionA.Red.Name, sampleNames(RGset)))
extensionA.Red[extensionA.Red.Name,] = getRed(RGset)[getProbeInfo(RGset,
type = "Control")[7,]$Address,]
extensionT.Red.Name =getProbeInfo(RGset, type = "Control")[8,]
$ExtendedType
extensionT.Red = matrix(NA_real_, ncol = ncol(RGset),
nrow = length(extensionT.Red.Name),
dimnames = list(extensionT.Red.Name, sampleNames(RGset)))
extensionT.Red[extensionT.Red.Name,] = getRed(RGset)[getProbeInfo(RGset,
type = "Control")[8,]$Address,]
extensionC.Green.Name = getProbeInfo(RGset, type = "Control")[9,]
$ExtendedType
extensionC.Green = matrix(NA_real_, ncol = ncol(RGset),
nrow = length(extensionC.Green.Name),
dimnames = list(extensionC.Green.Name, sampleNames(RGset)))
extensionC.Green[extensionC.Green.Name,] = getGreen(RGset)[getProbeInfo(RGset,
type = "Control")[9,]$Address,]
extensionG.Green.Name = getProbeInfo(RGset, type = "Control")[10,]
$ExtendedType
```

```
extensionG.Green = matrix(NA_real_, ncol = ncol(RGset),
nrow = length(extensionG.Green.Name),
 dimnames = list(extensionG.Green.Name, sampleNames(RGset)))
extensionG.Green[extensionG.Green.Name,] = getGreen(RGset)[getProbeInfo(RGset,
type = "Control")[10,]$Address,]


#HYBRIDISATION
hybridH.Green.Name = getProbeInfo(RGset, type = "Control")[11,]
$ExtendedType
hybridH.Green = matrix(NA_real_, ncol = ncol(RGset),
nrow = length(hybridH.Green.Name),
dimnames = list(hybridH.Green.Name, sampleNames(RGset)))
hybridH.Green[hybridH.Green.Name,] = getGreen(RGset)[getProbeInfo(RGset,
type = "Control")[11,]$Address,]
hybridM.Green.Name = getProbeInfo(RGset, type = "Control")[12,]
$ExtendedType
hybridM.Green = matrix(NA_real_, ncol = ncol(RGset),
nrow = length(hybridM.Green.Name),
                dimnames = list(hybridM.Green.Name, sampleNames(RGset)))
hybridM.Green[hybridM.Green.Name,] = getGreen(RGset)[getProbeInfo(RGset,
type = "Control")[12,]$Address,]
hybridL.Green.Name = getProbeInfo(RGset, type = "Control")[13,]
$ExtendedType
hybridL.Green = matrix(NA_real_, ncol = ncol(RGset),
nrow = length(hybridL.Green.Name),
dimnames = list(hybridL.Green.Name, sampleNames(RGset)))
hybridL.Green[hybridL.Green.Name,] = getGreen(RGset)[getProbeInfo(RGset,
type = "Control")[13,]$Address,]


#TARGET REMOVAL
target.Green.Name =getProbeInfo(RGset, type = "Control")[14:15,]
$ExtendedType
target.Green = matrix(NA_real_, ncol = ncol(RGset),
nrow = length(target.Green.Name),
dimnames = list(target.Green.Name, sampleNames(RGset)))
target.Green[target.Green.Name,] = getGreen(RGset)[getProbeInfo(RGset,
type = "Control")[14:15,]$Address,]


#Specificity I
specI.Green.Name =getProbeInfo(RGset, type = "Control")[32:34,]
$ExtendedType
specI.Green = matrix(NA_real_, ncol = ncol(RGset),
nrow = length(specI.Green.Name),
dimnames = list(specI.Green.Name, sampleNames(RGset)))
specI.Green[specI.Green.Name,] = getGreen(RGset)[getProbeInfo(RGset,
                            type = "Control")[32:34,]$Address,]
specI.Red.Name = getProbeInfo(RGset, type = "Control")[38:40,]
$ExtendedType
specI.Red = matrix(NA_real_, ncol = ncol(RGset),
nrow = length(specI.Red.Name),
dimnames = list(specI.Red.Name, sampleNames(RGset)))
```

```
specI.Red[specI.Red.Name,] = getRed(RGset)[getProbeInfo(RGset,
type = "Control")[38:40,]$Address,]

#Specificity II
specII.Red.Name = getProbeInfo(RGset, type = "Control")[44:46,]
$ExtendedType
specII.Red = matrix(NA_real_, ncol = ncol(RGset),
nrow = length(specII.Red.Name),
dimnames = list(specII.Red.Name, sampleNames(RGset)))
specII.Red[specII.Red.Name,] = getRed(RGset)[getProbeInfo(RGset,
type = "Control")[44:46,]$Address,]

#NON POLYMORPHIC
np.Red.Name =getProbeInfo(RGset, type = "Control")[47:48,]
$ExtendedType
np.Red = matrix(NA_real_, ncol = ncol(RGset),
nrow = length(np.Red.Name),
dimnames = list(np.Red.Name, sampleNames(RGset)))
np.Red[np.Red.Name,] = getRed(RGset)[getProbeInfo(RGset,
type = "Control")[47:48,]$Address,]
np.Green.Name =getProbeInfo(RGset, type = "Control")[49:50,]
$ExtendedType
np.Green = matrix(NA_real_, ncol = ncol(RGset),
nrow = length(np.Green.Name),
dimnames = list(np.Green.Name, sampleNames(RGset)))
np.Green[np.Green.Name,] = getGreen(RGset)[getProbeInfo(RGset,
type = "Control")[49:50,]$Address,]

#Normalisation
control=getProbeInfo(RGset, type = "Control")
normC.Green.Name=control[control[,2]=='NORM_C',4]
normC.Green = matrix(NA_real_, ncol = ncol(RGset),
nrow = length(normC.Green.Name),
dimnames = list(normC.Green.Name, sampleNames(RGset)))
normC.Green[normC.Green.Name,] = getGreen(RGset)
[control[control[,2]=='NORM_C',1],]
normG.Green.Name=control[control[,2]=='NORM_G',4]
normG.Green = matrix(NA_real_, ncol = ncol(RGset),
nrow = length(normG.Green.Name),
dimnames = list(normG.Green.Name, sampleNames(RGset)))
normG.Green[normG.Green.Name,] = getGreen(RGset)
[control[control[,2]=='NORM_G',1],]
normA.Red.Name=control[control[,2]=='NORM_A',4]
normA.Red = matrix(NA_real_, ncol = ncol(RGset),
nrow = length(normA.Red.Name),
dimnames = list(normA.Red.Name, sampleNames(RGset)))
normA.Red[normA.Red.Name,] = getRed(RGset)
[control[control[,2]=='NORM_A',1],]
normT.Red.Name=control[control[,2]=='NORM_T',4]
normT.Red = matrix(NA_real_, ncol = ncol(RGset),
nrow = length(normT.Red.Name),
```

```
dimnames = list(normT.Red.Name, sampleNames(RGset)))
normT.Red[normT.Red.Name,] = getRed(RGset)
[control[control[,2]=='NORM_T',1],]

#combine ctrl probe intensities
ctrl = rbind(as.matrix(BSCI.Green), as.matrix(BSCI.Red),
as.matrix(BSCII.Red), (stain.Red), (stain.Green),
(extensionA.Red), (extensionT.Red), (extensionC.Green),
(extensionG.Green), (hybridH.Green), (hybridM.Green),
(hybridL.Green),as.matrix(target.Green),as.matrix(specI.Green),
as.matrix(specI.Red), as.matrix(specII.Red),(np.Red[1,]),(np.Red[2,]),
(np.Green[1,]),(np.Green[2,]), as.matrix(normC.Green),
as.matrix(normG.Green), as.matrix(normA.Red), as.matrix(normT.Red))

#detection p-values
dp = detectionP(RGset, type = "m+u")

#add data for the new samples
if(exists("TypeII.Red.All")) {
TypeII.Red.All <- cbind(TypeII.Red.All,TypeII.Red)
TypeII.Green.All <- cbind(TypeII.Green.All,TypeII.Green)
TypeI.Red.M.All <- cbind(TypeI.Red.M.All,TypeI.Red.M)
TypeI.Red.U.All <- cbind(TypeI.Red.U.All,TypeI.Red.U)
TypeI.Green.M.All <- cbind(TypeI.Green.M.All,TypeI.Green.M)
TypeI.Green.U.All <- cbind(TypeI.Green.U.All,TypeI.Green.U)
ctrl.all <- rbind(ctrl.all, t(ctrl))
dp.all <- cbind(dp.all, dp)
}
else {
TypeII.Red.All <- TypeII.Red
TypeII.Green.All <- TypeII.Green
TypeI.Red.M.All <- TypeI.Red.M
TypeI.Red.U.All <- TypeI.Red.U
TypeI.Green.M.All <- TypeI.Green.M
TypeI.Green.U.All <- TypeI.Green.U
ctrl.all <- t(ctrl)
dp.all <- dp
}


#PCA of control-probe intensities
pca <- prcomp(na.omit(ctrl.all))
ctrlprobes.scores = pca$x
colnames(ctrlprobes.scores) = paste(colnames(ctrlprobes.scores),
'_cp', sep='')

save(TypeII.Red.All ,TypeII.Green.All ,TypeI.Red.M.All ,TypeI.Red.U.All ,
TypeI.Green.M.All ,TypeI.Green.U.All , file="../intensities.RData")
save(ctrl.all,ctrlprobes.scores, file = "../ctrlprobes.RData")
save(dp.all, file = "../detectionPvalue.RData")
```

```
# 2\_betasRev.R
# Sets NAs based on the detection p-value and calculates marker and
# sample call-rates. Filters samples based on sample calle-rate.
# Performs Quantile Normalisation and Calculates Beta-values.
# Originallly written by Benjamin Lehne (Imperial College London) and
# Alexander Drong (Oxford University)
# Modified by Hongmei Zhang (University of Memphis) with an
# application to a GEO dataset

require(limma)
anno=read.csv("/GPL13534_HumanMethylation450_15017482_v.1.1.csv",
     as.is=TRUE, skip = 7)
anno=anno[,c('Infinium_Design_Type','Color_Channel', 'CHR',
'MAPINFO', 'Name')]
cas=anno[substr(anno$Name, 1,3)=='ch.' & !(anno$CHR %in% c('X','Y')),]
cgs=anno[substr(anno$Name, 1,2)=='cg'& !(anno$CHR %in% c('X','Y')),]
auto = c(cgs$Name, cas$Name)
auto=as.matrix(auto)

#detection p-value
thres=1E-16
load('intensities.RData')
load('detectionPvalue.RData')
d=dp.all[rownames(TypeII.Green.All),colnames(TypeII.Green.All)]
TypeII.Green.All.d = ifelse(d<thres,TypeII.Green.All,NA)
TypeII.Red.All.d = ifelse(d<thres,TypeII.Red.All,NA)
d=dp.all[rownames(TypeI.Green.M.All),colnames(TypeI.Green.M.All)]
TypeI.Green.M.All.d = ifelse(d<thres,TypeI.Green.M.All,NA)
TypeI.Green.U.All.d = ifelse(d<thres,TypeI.Green.U.All,NA)
d=dp.all[rownames(TypeI.Red.M.All),colnames(TypeI.Red.M.All)]
TypeI.Red.M.All.d = ifelse(d<thres,TypeI.Red.M.All,NA)
TypeI.Red.U.All.d = ifelse(d<thres,TypeI.Red.U.All,NA)
rm(dp.all,d)

# autosomes -----------------------------------------------------------
samples=colnames(TypeI.Red.M.All)
category=auto
markers=as.matrix(intersect(rownames(TypeII.Green.All.d), category))
TypeII.Green = TypeII.Green.All.d[markers,samples]
TypeII.Red = TypeII.Red.All.d[markers,samples]
markers=intersect(rownames(TypeI.Green.M.All.d), category)
TypeI.Green.M = TypeI.Green.M.All.d[markers,samples]
TypeI.Green.U = TypeI.Green.U.All.d[markers,samples]
markers=intersect(rownames(TypeI.Red.M.All.d), category)
TypeI.Red.M = TypeI.Red.M.All.d[markers,samples]
TypeI.Red.U = TypeI.Red.U.All.d[markers,samples]

#raw betas
TypeII.betas = TypeII.Green/(TypeII.Red+TypeII.Green+100)
TypeI.Green.betas = TypeI.Green.M/(TypeI.Green.M+TypeI.Green.U+100)
TypeI.Red.betas = TypeI.Red.M/(TypeI.Red.M+TypeI.Red.U+100)
```

```
beta = as.matrix(rbind(TypeII.betas,TypeI.Green.betas,TypeI.Red.betas))
sample.call=colSums(!is.na(beta))/nrow(beta)
marker.call=rowSums(!is.na(beta))/ncol(beta)
#save(sample.call, marker.call, file='output/callRates.RData')
#save(beta, file='output/beta_raw.RData')

#call-rate filtering
callrate.thres=0.95
samples=names(sample.call[sample.call>callrate.thres])
markers=as.matrix(intersect(rownames(TypeII.Green.All.d), category))
TypeII.Green = TypeII.Green.All.d[markers,samples]
TypeII.Red = TypeII.Red.All.d[markers,samples]
markers=intersect(rownames(TypeI.Green.M.All.d), category)
TypeI.Green.M = TypeI.Green.M.All.d[markers,samples]
TypeI.Green.U = TypeI.Green.U.All.d[markers,samples]
markers=intersect(rownames(TypeI.Red.M.All.d), category)
TypeI.Red.M = TypeI.Red.M.All.d[markers,samples]
TypeI.Red.U = TypeI.Red.U.All.d[markers,samples]
rm(TypeII.Green.All.d,TypeII.Red.All.d,TypeI.Green.M.All.d,
TypeI.Green.U.All.d,TypeI.Red.M.All.d,TypeI.Red.U.All.d)

#QN
TypeII.Green=normalizeQuantiles(TypeII.Green)
TypeII.Red = normalizeQuantiles(TypeII.Red)
TypeI.Green.M = normalizeQuantiles(TypeI.Green.M)
TypeI.Green.U = normalizeQuantiles(TypeI.Green.U)
TypeI.Red.M = normalizeQuantiles(TypeI.Red.M)
TypeI.Red.U = normalizeQuantiles(TypeI.Red.U)
TypeII.betas = TypeII.Green/(TypeII.Red+TypeII.Green+100)
TypeI.Green.betas = TypeI.Green.M/(TypeI.Green.M+TypeI.Green.U+100)
TypeI.Red.betas = TypeI.Red.M/(TypeI.Red.M+TypeI.Red.U+100)
beta = as.matrix(rbind(TypeII.betas,TypeI.Green.betas,TypeI.Red.betas))
rm(TypeII.Green,TypeII.Red,TypeI.Green.M,TypeI.Green.U,
TypeI.Red.M,TypeI.Red.U, TypeII.betas,TypeI.Green.betas,TypeI.Red.betas)
save(beta, file="beta_QN.RData")
```

## CHAPTER 4

Screening genome-scale genetic and epigenetic data

In Chapters 2 and 3, we introduced different types of data at the genome scale. Given the feature of their high dimensionality, directly analyzing data at the genome scale with complex analytical methods may lead to substantially under-powered results. Depending on research goals, screening genome-scale data based on their associations with a phenotype or health outcome will enable us to target at potentially important genes or DNA methylation (CpG) sites. For instance, for studies with smoke exposure as a major risk factor, it has been shown that methylation is affected by some known factors such as smoking [82]. CpG sites not influenced by such exposure status potentially will only bring noise into subsequent analyses and thus reduce statistical power. On the other hand, CpGs with DNA methylation not associated with a health outcome may not be of great interest either. Screening genetic and epigenetic variables or features at the genome scale has become an overwhelmingly important step in multiple fields of genetic and epigenetic studies such as cancer, obesity, and allergic diseases.

We focus on screening methods that are built upon associations with some post-hoc adjustment for multiple comparisons. We denote factors, such as smoke exposure status or a health outcome, used to exclude potentially unimportant genetic or epigenetic variables as "screening factors". Most commonly used screening approaches built upon associations are through regressions. In this chapter, we introduce two directions of development in this filed; one mainly relies on training and testing data using the concept of cross validation, and the other one mainly focuses on correlations.

4.1 SCREENING VIA TRAINING AND TESTING SAMPLES

When screening factors in regressions are included as independent variables, a primary limitation lies in effectively controlling for multiple testing. Two popular adjustment methods are the Bonferroni-based method [31, 32] and the Benjamini-Hochberg method for controlling the false discovery rate (FDR) [66, 7]. These methods alter the p-value or critical value to control for type I error. Bonferroni correction is the most conservative approach. An adjusted p-value is calculated by multiplying the linear regression p-value by the total number of comparisons (m) and variables with adjusted p-values less than the significance level α are treated as being statistically significant. The FDR method first orders the p-values, $P(k)$ for $k \in 1...m$, such that lower ordered p-values that are less than or equal to $k/m \times \alpha$ lead to rejection of the null hypothsis [7]. It follows that the conservative Bonferroni-based method is not able to effectively control type II error while the FDR-based method cannot control type I error as desired.

In this section, we introduce a recently developed method that has the potential to control both types I and II errors. It utilizes existing statistical techniques to screen genome-scale variables. The method is built upon training and testing data. Even with data generated under the same mechanisms in the same population, associations between genetic and epigenetic variables and screening factors may vary from one data set to another, which directly impacts type I error rates, and indirectly leads to a loss of statistical power in subsequent analyses. Using training and testing data will improve the reproducibility of selected variables, and earn the ability to control for both types I and II errors [118].

This approach starts from randomly dividing the original data into two parts as training data and testing data, respectively. Following literature suggestions, in general, 2/3 of the data are included in the training data set to maximize statistical power.[27] Next, linear regressions are applied to the training data to calculate the p-values for the association between a variable (dependent variable) and screening factors (independent variables), for instance, DNA methylation at a CpG site and smoke exposure status. Ordinary least squares or robust regressions can be applied to infer the parameters. Compared to linear regressions, robust regressions have more relaxed assumptions about normality and presence of outliers in the data. A variable, e.g., a CpG site, is included as a candidate variable if the screen factor(s) of interest is statistically significant according to a pre-specified significance level, e.g., 0.05. A set of candidate variables are then selected by use of training data.

The candidate variables are further tested using the testing data with linear regressions. For one pair of training and testing data sets, a candidate variable is deemed as being important if the significance still holds in the testing data. This screening process will be repeated for a certain number of times, and at each time a pool of candidate variables are selected based on one randomly selected set of training and testing data.

The screening process is summarized in Figure 4.1 (the solid arrows). A variable is treated as being potentially important if it is selected by at least $m\%$ of the randomly selected sets of training and testing data. The value of m is pre-specified. A suggestion of choosing m has been given in the literature [118]. Basically, taking m close to 50 works the best with significance level of 0.05 for both the training and testing steps. If a higher significance level is chosen in the testing step, then a higher value for cutoff percentage should be used.

The R package, `ttScreening`, can be used to implement the above approach. It is originally designed for epigenetic data, in particular, DNA methylation data. The package can be applied to other genetic or epigenetic data as long as their measurements can be treated as being continuous. Two data sets are used as input. One data set is at the genome scale such that each row represents measures at one locus across all subjects. The other data set is composed of measurements of one or more risk factors or health outcomes. Since this approach is closely related to the method introduced in the next section, we hold our discussions on the implementation of this package till then.

4.2 SCREENING INCORPORATING SURROGATE VARIABLES

Implementation of training and testing data introduces a potential to control both types I and II errors [118]. However, other issues arise when performing screening based on associations. Variations in data can be explained by some known factors but also by some unknown factors [93]. Accounting for variations introduced by unknown factors will potentially improve screening quality and efficiency. The method discussed below is an extension of that introduced in Section 4.1. It incorporates surrogate variable analysis [93], which identifies unknown latent variables, in conjunction with the training and testing approach noted earlier [27].

This approach consists of two consecutive components, surrogate variable analysis followed by utilization of the screening method discussed in Section 4.1. Surrogate variable analysis (SVA) aims to identify and estimate latent factors or surrogate variables (SVs) that potentially affect the association of a variable with a screening factor or screening factors, e.g., DNA methylation (variable) and smoke exposure status (a known screening factor) [93]. Including surrogate variables into the screening process has the potential to reduce unexplained variations in the variable, adjust for confounding effects, consequently, improve the accuracy of screening in terms of importance of variables that pass the screening [93].

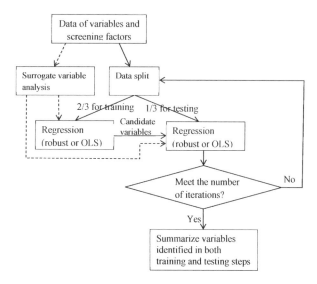

Figure 4.1 The flow chart of the screening algorithm that involves surrogate variables, adapted from Ray et al. [118].

Surrogate variables are inferred prior to screening using the method developed by Leek and Storey [93] (Figure 4.1, dashed arrows). These SVs are developed by removing the amount of variation in the variables due to screening factors and then decomposing the remaining residuals to identify an orthogonal basis of singular vectors that can be reproduced. These vectors are further examined for significant variation to form surrogate variables. Leek and Storey built an R package to perform

SVA. The first step in SVA is to identify the number of surrogate variables based on the data using one of two methods, an approach based on a permutation procedure originally proposed by Buja and Eyuboglu in 1992 [15], and the other provides an interface to the asymptotic approach proposed by Leek in 2011 ([91]). Once the number of surrogate variables is calculated, they are then estimated using one of three algorithms, the iteratively re-weighted, supervised, or two-step method. The iteratively re-weighted method is for empirical estimation of unknown factors, the supervised method is for the situation that control probes (i.e., probes such that genes are unlikely to be differential expressed) are known, and the two-step method estimates surrogate variables based on the subset of rows affected by unmodeled dependence [92]. Using this approach, conditional on the data, a number of latent unknown variables will be identified and estimated. These latent variables along with the known screening factors will be included in the training-and-testing-based method to detect potentially important variables. This approach has been built into the ttScreening.

In the following, we apply the ttScreening package to a simulated data set which has $2,000$ variables and 2 observable screening factors with measures from 300 subjects. Of these $2,000$ variables, the first 100 are important variables. We start from simulating a data set with the settings noted above with the two variables simulated from normal distributions using rmvnorm.

```
> library(mvtnorm)
> # 300 subjects, the first 100 are important variables,
+    and 2000 candidate variables
> nsub = 300
> imp = 100
> num = 2000
> set.seed(1)
> # two observable screening factors
> meanx = c(1,1)
> Covx = diag(c(1,1))
> x = rmvnorm(nsub, meanx, Covx)
```

In addition, we simulate five "latent" screening factors, which will be included in the generation of the $2,000$ variables.

```
> # five latent screening factors
> surMean = c(0,3,0,2,0)
> CovSur = diag(rep(0.5,5))
```

```
> sur = rmvnorm(nsub, surMean, CovSur)
```

As for the association of variables with the screening factors x, we consider three settings such that some variables will be associated with both screening factors (Design1 and Design2) but others only associated with the "latent" screening factors (DesignSur).

```
> Design1 = as.matrix(cbind(rep(1,nsub),x,x[,1]*x[,2],sur))
> Design2 = as.matrix(cbind(rep(1,nsub),x[,1],x[,1]*x[,2],sur))
> DesignSur = as.matrix(sur)
```

The following several lines specify regression coefficients corresponding to each design,

```
> beta = c(0.5, rep(0.4,3))
> sbeta = rmvnorm(1,rep(0.5,5),diag(rep(0.01,5)))
>
> # beta matrix defining different associations with
> # observable and latent screening factors
> betaDesign1 = as.matrix(c(beta,sbeta))
> betaDesign2 = as.matrix(c(beta[c(1,2,4)],sbeta))
> betaSur = as.matrix(t(sbeta))
```

Finally, with the design matrices and regression coefficients specified, data, y, for the 2,000 variables are then simulated using a multivariate normal distribution.

```
> imp1.mu = matrix(Design1%*%betaDesign1,nrow = nsub,ncol = (imp*0.9))
> imp2.mu = matrix(Design2%*%betaDesign2,nrow = nsub,ncol = (imp*0.1))
> noimp.mu = matrix(DesignSur%*%betaSur,nrow = nsub,ncol = num-imp)
> mu.matrix = cbind(imp1.mu, imp2.mu, noimp.mu)
> CovErr = diag(rep(0.5,num))
> error = rmvnorm(nsub,mean = rep(0,num),sigma = CovErr,method = "chol")
> y = t(mu.matrix+error)
```

We next apply ttScreening to screen variables based the interaction between the two screening factors in x. First, we only utilize training and testing data without inclusion of surrogate variables.

```
> runs = ttScreening(y = y, formula = ~x[,1]+x[,2]+x[,1]:x[,2],
+ imp.var = 3, data = data.frame(x),
+ B.values = FALSE, iterations = 100, cv.cutoff = 50,
+ n.sv = 0, train.alpha = 0.05, test.alpha = 0.05,
+ FDR.alpha = 0.05, Bon.alpha = 0.05, percent = (2/3),
+ linear = "ls", vfilter = NULL, B = 5,
+ numSVmethod = "be", rowname = NULL,maxit = 20)
```

```
[1] "No surrogate variables used in the analysis"
>
> length(runs$TT.output[,1])
[1] 95
> length(runs$FDR.output[,1])
[1] 103
> length(runs$Bon.output[,1])
[1] 84
> sum(runs$TT.output[,1]%in% seq(1,imp))
[1] 95
> sum(runs$FDR.output[,1]%in% seq(1,imp))
[1] 100
> sum(runs$Bon.output[,1]%in% seq(1,imp))
[1] 84
```

In the above function, ordinary least square regressions, set by linear = "ls", are used for the screening, y is the genome-scale data, x[,1] and x[,2] represent two screening factors, and imp.var=3 denotes that the screening is with respect to the interaction effects between the two factors. We can also choose to screen variables based on both x[,1] and x[,2], in which case we set imp.var=c(1,2).

The argument iteration sets the number of times to repeat the training and testing procedure and cv.cutoff is used to set the minimum frequency required for a variable to be treated as an important variable. The higher the frequency, the more likely the variable is informative. The default value of cv.cutoff is 50. The two arguments train.alpha and test.alpha define the statistical significance levels used in the training and testing data, respectively, and their default values are both 0.05. We use percent to define the size of training data, and, in the above example, data from 2/3 of the subjects will be used.

Finally, argument n.sv=0 is used to indicate that the screening will not include any surrogate variables but only observable screening factors, i.e., x[,1] and x[,2]. Selected variables will be stored in the object TT.output along with frequencies for each selected variable. The first column of TT.output contains the variable names that pass the screening. In this example, 95 variables are selection and all these 95 variables are among the 100 important variables. The ttScreening package also provides users with access to other screening methods: FDR- and Bonferroni-based methods, and output can be accessed via FDR.output and Bon.output, respectively, with similar structure compared to TT.output (except for selection frequencies). The FDR-based approach selected 103 variables with 3 false positive selections.

The Bonferroni approach selected 84 and all of them are among the 100 important ones.

To include surrogate variables in the screening process, we set n.sv=NULL. The argument numSVmethod is to specify a method used to determine the number of surrogate variables, and "be" is the method by Buja and Eyuboglu [15]. The other approach is proposed by Leek and labeled as "leek" [91]. To specify the method used to estimate surrogate variables, sva.method is used, and in this example, the two-step approach is selected. The specifications of the other two approaches are "irw" for the iteratively re-weighted approach, and "supervised" for the supervised approach.

```
> library(ttScreening)
> runs = ttScreening(y = y, formula = ~x[,1]+x[,2]+x[,1]:x[,2],
+ imp.var = 3, data = data.frame(x), sva.method = "two-step",
+ B.values = FALSE, iterations = 100, cv.cutoff = 50,
+ n.sv = NULL, train.alpha = 0.05, test.alpha = 0.05,
+ FDR.alpha = 0.05, Bon.alpha = 0.05, percent = (2/3),
+ linear = "ls", vfilter = NULL, B = 5,
+ numSVmethod = "be", rowname = NULL,maxit = 20)
Number of significant surrogate variables is:  1
[1] "1 surrogate variables used in the analysis"
>
> length(runs$TT.output[,1])
[1] 101
> length(runs$FDR.output[,1])
[1] 113
> length(runs$Bon.output[,1])
[1] 100
> sum(runs$TT.output[,1]%in% seq(1,imp))
[1] 100
> sum(runs$FDR.output[,1]%in% seq(1,imp))
[1] 100
> sum(runs$Bon.output[,1]%in% seq(1,imp))
[1] 100
```

With the inclusion of surrogate variables, the results are improved for the method in ttScreening and the Bonferroni approach, especially the latter. False positive selections of FDR-based approach gets more severe.

4.3 SURE INDEPENDENCE SCREENING

This approach treats screening factors as dependent variables, and all variables to be screened are included in a regression model as independent variables. In this case, the screening is built upon variable selections. The method is proposed by Fan and Lv [38], and is composed of two steps. The first step uses the concept of sure independence screening, also known as correlation learning [45], to filter out the variables that have weak correlations with the screening factor. A property of sure screening is that all the important variables will be kept after variable screening with probability approaching to 1. Sure independence screening is then followed by implementation of variable selection techniques, such as the SCAD method proposed by Fan and Li [37], the Adaptive LASSO by Zou [165], the Dantzig selector by Candes and Tao [16], or other Bayesian methods [52, 159].

4.3.1 Correlation learning

The sure independence screening method is built upon correlation learning. Correlation learning is an old and computationally simple approach for variable selection. It starts from a set of simple linear regressions with predictors (variables) standardized and the response variable (screening factor) centered. Following the notations given by Fan and Lv [38], estimates of regression coefficients in these simple linear regressions are defined as $\tilde{\beta} = X^T y / n$, where X is the $n \times p$ data matrix of variables with n the sample size and p the number of variables, and $y = (Y_1, \cdots, Y_n)^T$ is the centered screening factor. It is not difficult to see that $X^T y$ is a vector of sample correlations between each of the variables and the screening factor, re-scaled by the standard deviation of y.

Let $\mathcal{M}_\star$ denote the true model with size $s = |\mathcal{M}_\star|$, and $\mathcal{M}_\gamma$ denote a submodel with size $[\gamma n]$, where γ is a tuning parameter that needs to pre-specify and $[\gamma n]$ denotes the integer part of γn. Applying the sure independence screening, variables in the submodel $\mathcal{M}_\gamma$ are chosen based on $|\omega_i| = n|\tilde{\beta}_i|$ such that the top $[\gamma n]$ variables are selected, and included in the next step of screening. Fan and Lv showed that

$$P(\mathcal{M}_\star \subset \mathcal{M}_\gamma) \to 1, \quad \text{as } n \to \infty,$$

for some given γ. Under some regulatory conditions, it is suggested taking $[\gamma n] = n - 1$ or $[\gamma n] = n / \log(n)$.

With variables to be screened in the regressions as independent variables, there are some concerns in the selection of truly important variables using the sure independence screening approach. Unimportant variables highly correlated with the important ones may be selected. In general, collinearity adds in difficulty to correctly select truly important variables. Also, a variable that is marginally uncorrelated but jointly correlated with the screening factor will not be selected. To overcome these problems, Fan and Lv proposed two solutions. One approach iteratively applies the sure independence screening approach, and the other approach groups and transforms the variables aiming to improve the chance of selecting the correct variables and reduce collinearity.

Iterative correlation learning   Iterative correlation learning is to apply the correlation learning approach iteratively to variable selections. This approach first selects a subset of variables using the sure independence screening followed by variable selection techniques such as SCAD or LASSO based on joint information of $[n/\log(n)]$ variables chosen from the correlation learning. Denote this set of selected variables as $\mathcal{A}_1$. Next, the residuals are calculated by regressing the screening variable y over the selected variables. The residuals are then treated as new responses, and are used in the next step of sure independence screening followed by variable selections over the set of variables not in $\mathcal{A}_1$. The set of selected variables is denoted as $\mathcal{A}_2$, which is mutually exclusive with $\mathcal{A}_1$. The residuals are uncorrelated with variables in $\mathcal{A}_1$, and thus the selection in this step will be less influenced by high correlations between unimportant and important variables. The strength of using residuals will fulfill the goal of selecting important variables that are marginally uncorrelated but jointly correlated with the response. However, if variable A is highly correlated with a variable of interest, then variable A may still want to be selected. Using residuals as the response may exclude such type of variables.

Nevertheless, the process of sure independence screening followed by variable selection continues and each time we obtain a set of selected variables disjoint with previous sets of selected variables. The final set of variables that pass the screening is the union of all the disjoint sets

$$\mathcal{A} = \bigcup_{i=1}^{l} \mathcal{A}_i,$$

where l can be chosen such that $|\mathcal{A}| = d < n$, as suggested by Fan and Lv [38].

Variable grouping and transformation   The goal of grouping variables is to utilize joint information from a group of variables. The sure independence screening approach is applied to the groups, which are then included in subsequent variable selections. However, this approach may not be able to solve the concerns on correlations and collinearity. Transforming the variables is likely to be more meaningful. Fan and Lv [38] suggested two approaches: subject-related transformation and statistical transformation. Subject-related transformations are comparable to the transformations commonly used in statistical analyses, e.g., centering variable, or using differences in variables, to reduce correlations or collinearity. Statistical transformations utilize principal components analyses to identify weakly correlated new variables, on which the sure independence screening are applied to select variables. One disadvantage of this approach is the interpretation of selected variables. In genetic and epigenetic related studies, each component may cover different genes on different chromosomes and it may be hard to summarize the functionality of identified components, and subsequent selection results.

To implement the approach of sure independence screening coupled with variable selections, Saldana and Feng [123] developed an R package built upon the work of Fan and Lv [38]. The SIS function in the package implements the iterative sure independence screening approach, and then uses the R packages ncvreg and glmnet for variable selections using SCAD, minimax concave penalty (MCP), or LASSO.

```
> model = SIS(x, y, family = "gaussian",
penalty = "lasso",tune = "bic")
> model$ix
```

In the above, x provides the design matrix for p variables with n rows and p columns, and y is the response vector (for the screening variable) with length of n. The argument family refers to the assumed family of distributions of data. In addition to gaussian, other choices of family are binomial for binary response variable and cox for time to events data. If family="cox", then y should be an object from Surv() in the package survival. Penalty is utilized in variable selections to exclude variables that do not contribute to the explanation of variations in y. The default penalty is SCAD. A tuning parameter is critical in variable selections. Besides bic, other tuning parameter selectors include, "ebic" (Extended Bayesian Information Criteria,[19]), "aic", and "cv" (cross validation). The indices of the selected variables is included in the vector ix.

4.4 NON- AND SEMI-PARAMETRIC SCREENING TECHNIQUES

In addition to the parametric methods, non-parametric and semi-parametric approaches have been proposed to screen variables. In this section, we discuss two types of such approaches, random forest-based methods and support vector machine.

4.4.1 Random forest

Random forest [12] is a machine learning approach for classification or regression. It constructs a large number of decision trees or classifiers and aggregates the results from the trees to reach a final conclusion on classifications and regressions. The approach is built upon a successful machine learning technique, bagging [11]. Each tree in bagging is independently constructed using bootstrap samples, and final decision is reached by a majority vote.

Random forest, compared to bagging, adds another layer to the construction of trees. For each bootstrap sample, instead of building trees using all the variables, in random forest, each tree is built upon a subset of variables. That is, at each node of a tree, the determination of splitting is based on a subset of randomly selected variables instead of using all the variables. An immediate advantage is the computational efficiency. More importantly, this strategy is robust against over fitting and performs well compared to other machine learning approaches such as support vector machines to be discussed later.

An R package `randomForest` is available to implement the algorithm proposed by Breiman [12]. Liaw et al. [98] discuss the usage and features of the package, and provide insightful notes for practical application of the package. In the following, we outline the algorithm, followed by examples of using randomForest.

The algorithm in randomForest   The algorithm starts from drawing a random bootstrap sample from the original data.

1. Draw a random subset of data from the complete data.

2. At each node, randomly select a subset of variables, and determine the best split based on those variables. Do these for all the l nodes, where l is the maximum number of nodes allowed.

3. Repeat steps 1. and 2. for a number of bootstrap samples to build a forest.

4. Use the rules of each tree in the forest to predict the outcome of the remaining data. The remaining data are not used in the construction of the forest, and thus are called out-of-bag (OOB) data.

5. A prediction of the OOB data is made by aggregating across the forest. It is majority votes across all the trees for classifications and average across all tree outcomes for regressions. An OOB prediction error rate is calculated for the OOB data. For classifications, % of misclassification is the prediction error, and for regressions, mean squared errors are calculated by comparing the observed and the predicted values.

The `randomForest` package provides an important statistics, named "variable importance" that can be used to select important variables. Variable importance is evaluated based on a variable's contribution to reduce OOB prediction error rates, and estimated based on permutating the values of the variables of interest. The above algorithm is implemented in the function `randomForest` of the package. We use the Forensic Glass data set in Ripley [119], which is applied in Liaw et al. [98], to demonstrate this function. The data has 9 measured physical characteristics for each of 214 glass fragments and these fragement are classified into 6 categories (noted as `type` in the codes below).

```
> library(randomForest)
> library(MASS)
> data(fgl)
> head(fgl)
> set.seed(17)
> RFobject = randomForest(type ~ ., data = fgl,
+            mtry = 2, importance = TRUE,
+            do.trace = 100)
```

The data format is as the following,

```
> head(fgl)
     RI    Na   Mg   Al    Si    K   Ca Ba   Fe type
1  3.01 13.64 4.49 1.10 71.78 0.06 8.75  0 0.00 WinF
2 -0.39 13.89 3.60 1.36 72.73 0.48 7.83  0 0.00 WinF
3 -1.82 13.53 3.55 1.54 72.99 0.39 7.78  0 0.00 WinF
4 -0.34 13.21 3.69 1.29 72.61 0.57 8.22  0 0.00 WinF
5 -0.58 13.27 3.62 1.24 73.08 0.55 8.07  0 0.00 WinF
6 -2.04 12.79 3.61 1.62 72.97 0.64 8.07  0 0.26 WinF
```

When building the forests, `mtry=2` variables are selected from the 9 available ones and used in the determination of split at each node. The default values for `mtry` is $\sqrt{p}$ for classifications, and $p/3$ for regressions. By setting `importance=TRUE`, importance of each predictor will be estimated, and setting `do.trace=100`, OOB errors are printed for every 100 trees. The importance values will help select important variables for subsequent analyses, as demonstrated below.

```
> GiniSorted = sort(RFobject$importance[,8], dec = TRUE)
> Labels = names(GiniSorted)
> plot(GiniSorted, type = "h",xlab = "Physical characteristics",
xaxt = "n",main = "Decrease in mean Gini indices")
> axis(side = 1, at = seq(1,9), labels=Labels)
```

In the above codes, `RFobject$importance` is a matrix composed of importance values for each physical characteristic across all the 6 categories as well as averaged importance values. The 8-th column of the matrix is the average Gini indices, which can be used to determine which characteristic could be removed from further analyses for any of the 6 categories. To visualize the pattern, we sort the average Gini indices using the function `sort`, and the last two lines above generate the graph shown in Figure 4.2. Using the concept of scree plot introduced in Chapter 2, *Ba* and *Fe* are likely the least important variables.

Recursive random forest. Applications of random forest in variable screening have been discussed in various studies [29, 47]. Measures of variable importance from the random forest output give users some idea about the importance of each variable. However, they can be used more effectively. Granitto et al. [61] and Díaz-Uriarte and De Andres [26] further independently utilize variable importance, along with OOB error rates, and include them in an iterative process of screening important variables. The advantage of iteratively implementing random forest is the potential of improving stability of variables that pass the screening, and consequently being beneficial to convincible biological interpretability of results obtained. The approach of Granitto et al. [61] and that of Díaz-Uriarte and De Andres [26] are comparable. In the following, we focus on the approach of the later [26].

The procedure of recursive random forest starts from running one random forest will all the candidate variables. We then extract variable importance measures from the random forest outputs. After sorting the importance measures, variables with importance measures at the lower

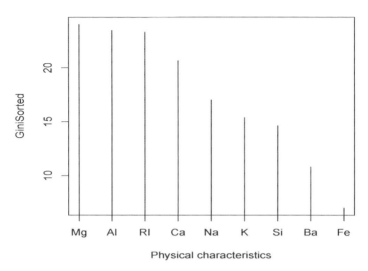

Figure 4.2 Sorted mean Gini indices of each physical characteristic from randomForest (the Forensic Glass data).

end are dropped. The percentage of variables to be dropped is a sub-jective selection and controlled by the aggressiveness of the user. The random forest procedure is then run again based on the subset of the variables. The process continues until the OOB error rates meet a pre-specified cut off, then the variables from the prior step becomes the selected variables to be used in subsequent analyses (Figure 4.3). Two criteria are proposed to specify the cut off that stops the iterations [26]. The first one is to choose the smallest number of variables whose error rate is within one standard errors of the minimum error rate of all forests, or selecting the set of variables that give the smallest error rate. Based on the findings noted by Díaz-Uriarte and De Andres [26], using the one standard error rule tends to choose smaller numbers of variables. In the diagram 4.3, the rule of minimum error rate is employed.

4.4.2 Support vector machine

Support vector machine (SVM) is a classification algorithm and it finds a separating hyperplane with the maximal margin between two classes of data determined by support vectors [143]. SVM is originally designed as

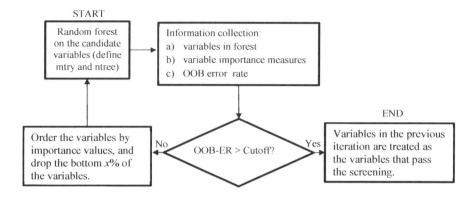

Figure 4.3 The flow chart of the recursive random forest screening algorithm [26].

a classifier for binary outcomes, and has been extended to deal with multiclass problems. Some studies compared between random forest (RF) and SVM and suggested that RF can outperform SVM for specific type of data [101], but in bioinformatics studies, it seemed SVM has been recommended over RF [136]. Gaussian kernels are commonly applied in SVM to flexibly classifying the objects. In addition to be applied as classifiers, the weights of an SVM classifier can be used to estimate importance values for each variable. The fit function in the package rminer is able to estimate such importance values, based on which we screen variables.

```
> library(rminer)
> M = fit(type ~ ., data = fgl, model = "svm", kpar = list(sigma = 0.10))
> svmImp = Importance(M, data = fgl, method = "1D-SA")
> Importance = svmImp$imp[1:9]
> names(Importance) = colnames(fgl)[1:9]
> sortImp = sort(Importance,  dec = TRUE)
> Labels = names(sortImp)
> plot(sortImp, type = "h",xlab = "Physical characteristics",xaxt = "n",
+   main = "Decrease in mean Gini indices")
> axis(side = 1, at = seq(1,9), labels = Labels)
```

We first run the svm classifier using the function fit. Many other choices for model are available, e.g., bagging and lm for regular linear regressions. A Gaussian kernel is used and the tuning parameter in the kernel is specified by kpar, which is 0.10. The importance values are then determined using sensitivity analyses in Importance. Several sensitivity analyses methods are available and the default approach is 1D-SA, a one-dimensional sensitivity analysis not considering interactions. Once the importance values are inferred (svmImp$imp), we plot the importance values (Figure 4.4) and can use the scree plots introduced in chapter 2 to determine variables potentially important.

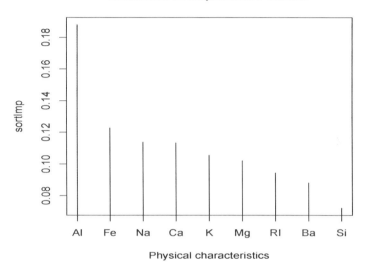

Figure 4.4 Sorted importance values of each physical characteristics from SVM in fit (the Forensic Glass data).

## CHAPTER 5

Cluster Analysis in Data mining

In general, data mining based on cluster analyses is to visualize the patterns in the data and is descriptive, although hypotheses testing on cluster profile differentiation is usually conducted as well. We introduce commonly used non-parametric and parametric clustering methods. Non-parametric clustering approaches are in general unsupervised learning and include methods based on data partitioning and methods aiming to cluster the data hierarchically. Partitioning-based methods aim to maximize certain criteria to map vectors to clusters, while hierarchical clustering methods construct "trees" showing similarities between groups. Parametric approaches in general model the distribution of data as a mixture of a certain number of distributions such that each part of the data is assumed to be generated from one of the mixtures. The goal is to optimize the distribution mixture to determine the number of clusters. Thinking of data matrix with n rows (representing n subjects) and m columns, in both non-parametric and parametric clustering methods, the objects to be clustered (clustering objects) can be rows (subjects) or columns (variables) determined by research interest.

5.1   NON-PARAMETRIC CLUSTER ANALYSIS METHODS

Non-parametric cluster analyses identify clusters based on distance between clustering objects. Various ways have been proposed to define distances, among which distance for continuous variables based on Pearson sample correlations, Euclidean distance, and Manhattan distance are used often. Gower distances can be applied to a mixture of continuous and categorical variables. In the following, we introduce each of these distance measures.

5.1.1 Distances

Let $\boldsymbol{x} = \{x_1, \cdots, x_n\}, \boldsymbol{y} = \{y_1, \cdots, y_n\}$ denote two vectors.

A Pearson sample correlation-based distance is

$$d_{cor}(\mathbf{x}, \mathbf{y}) = 1 - \frac{\sum_{i=1}^{n}(x_i - \bar{x})(y_i - \bar{y})}{\sqrt{\sum_i(x_i - \bar{x})^2 \sum_i(y_i - \bar{y})^2}} = 1 - r(\mathbf{x}, \mathbf{y}),$$

with zero distance represented by a perfect correlation between $\boldsymbol{x}$ and $\boldsymbol{y}$. In the Euclidean metric, the distance between $\boldsymbol{x}$ and $\boldsymbol{y}$ is defined as

$$d_{euc}(\mathbf{x}, \mathbf{y}) = \sqrt{\sum_{i=1}^{n}(x_i - y_i)^2},$$

These two distance metrics, d_{cor} and d_{euc}, are closely related. Denote by $\boldsymbol{z}_x$ and $\boldsymbol{z}_y$ standardized $\boldsymbol{x}$ and $\boldsymbol{y}$, respectively. We have

$$d_{euc}(\mathbf{z}_x, \mathbf{z}_y) = \sqrt{2n \, d_{cor}(\mathbf{z}_x, \mathbf{z}_y)}.$$

To reduce the impact of outliers, sometimes Spearman sample correlation-based distances are also used. Spearman correlations are calculated in the same way as in Pearson correlations except that ranks of data instead of original data are used to determine the correlations.

Manhattan distance is in the L_1 norm and defined as

$$d_{man}(\mathbf{x}, \mathbf{y}) = \sum_{i=1}^{n}|x_i - y_i|,$$

Unlike the Pearson or Spearman sample correlation-based distance metric, the Euclidean and Manhattan distance metrics do not take into account variations in data. The Mahalanobis distance metric, on the other hand, incorporates data variations into its distance definition, but assumes $\boldsymbol{x}$ and in $\boldsymbol{y}$ are from the same multivariate distribution with mean μ and covariance Σ,

$$d(\mathbf{x}, \mathbf{y})_{mah} = (\mathbf{x} - \mathbf{y})^T \Sigma^{-1} (\mathbf{x} - \mathbf{y}).$$

The Mahalanobis distance is equivalent to the Euclidean distance when Σ is the identity matrix.

In many situations, variables in cluster analyses are a mixture of continuous and categorical variables. In this case, Gower distance can be applied [60], which is an average of distances across all the variables with each distance between 0 and 1. For binary variables, Gower distance is calculated as proportion of mismatching, defined based on Jaccard similarity. Similar calculations are applied to nominal variables. For continuous variables, the distance is calculated as $|x_i - y_i|/R_i$ with R_i denoting range of variable i for all clustering objects. For ordinal variables, ranking the variables is first applied and then the distance is calculated in the same way as for continuous variables. In some other situations, the interest is in the distance between distributions. Metrics that assess such distances include the Kullback-Leibler distance (KLD) and its special case mutual information (MI). The KLD metric measures the difference between two probability distributions. MI is KLD and measures the distance between joint probability distribution of two random variables and product of their marginal distributions. Essentially, MI measures distance from independence.

With distance metric defined, clustering objects are included into a group if they are close to each other defined by their distance to a cluster center or distances between each objects. Through simulated data, we introduce three commonly applied clustering techniques: partitioning-based methods, hierarchical clustering methods, and a hybrid of these two approaches. This is then followed by a demonstration using gene expression data.

5.1.2 Partitioning-based methods

The K-means approach is among the most used partitioning-based clustering techniques due to its simplicity. The concept was first introduced by Cox [22], and the method later experienced various improvements [102, 41, 105, 64]. The algorithm by Hartigan and Wong [64] is used most often. It starts from specifying the number of clusters, K, and initial centroid of each cluster. Let c with length K denote a collection of centroids, $c = \{c_1, \cdots, c_k, \cdots, c_K\}$ with $c_k = \{c_{k1}, \cdots, c_{kp}\}$ being the centroid of cluster k for p components. Depending on the clustering objects, the p components can be p variables or p subjects. After selecting a distance metric, we assign each data point to a specific cluster following the steps below:

1. Calculate the distance between each data point and cluster centers. The Euclidean distances are commonly used.

2. Each data point is assigned to a cluster based on $argmin_{c_k \in c} d(c_k, x)$, where x denotes a data point with length p (p components).

3. Update the centroid of each cluster, which is calculated as the mean of the data points in each cluster.

4. If no data point is reassigned to a different cluster, then stop. Otherwise, repeat steps 1 through 3.

To determine the number of clusters, we can choose K by optimizing the total within-cluster variation. for continuous variables, it is defined as

$$SS_{within} = \sum_{x_i \in C_k} (x_i - c_k)^T (x_i - c_k),$$

where C_k denotes cluster k, and x_i the ith data point. The number of clusters can be determined by an empirical rule, the so-called "elbow" rule; plotting a set of SS_{within}'s sorted by the number of clusters and selecting K such that, at that K, SS_{within} experiences the biggest reduction followed by a flattened out SS_{within} reduction. This is essentially the scree test commonly used in factor analyses [69]. In Figure 5.1, the biggest reduction of SS_{within} happens at 3 clusters, and then the reduction of SS_{within} is quite small between 4 and 3 clusters. Following the "elbow" rule, with 3 clusters, the data are likely to be grouped neatly.

The advantage of the K-means approach exists in its easy to follow algorithm and clear separation of data for data with roughly regular shape in distributions. However, this approach is sensitive to the scales of data and works better if data distributions are with spherical or elliptical shapes. In addition, K-means is sensitive to outliers. More robust variants of Hartigan and Wong's method [64] have been proposed, such as generalized K-means and trimmed K-means [44], the improved K-means via outlier removal [65], among others [68, 53]. Even the case, the classical algorithm by Hartigan and Wong in general does well and is still mostly used.

The `kmeans` function in R implements the K-means approach. We demonstrate this through a simulated data set:

```
> set.seed(12345)
> x1 = c(rnorm(15, sd = .05), rnorm(20, mean = 1, sd = .05),
+        rnorm(25,mean = 1.5, sd = .05))
> x2 = c(rnorm(15,mean = 3, sd = 2), rnorm(20, mean = 11, sd = 1),
+        rnorm(25,mean = 12, sd = 1))
> x = cbind(x1, x2)
```

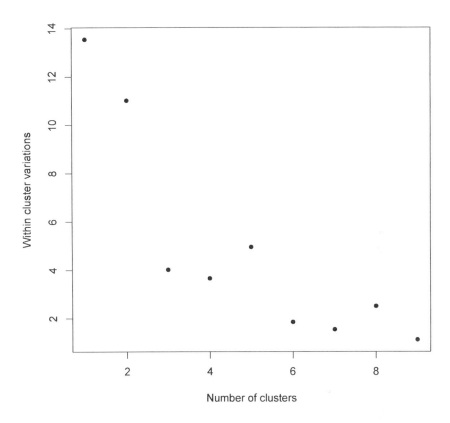

Figure 5.1   Illustration of utilizing the "elbow" rule to determine the optimized number of clusters.

```
> rownames(x) = c(rep(1,15), rep(2,20), rep(3,25))
> plot(x, pch = as.character(rownames(x)), asp = 1)
> x1Sd = x1/sd(x1)
> x2Sd = x2/sd(x2)
> x = cbind(x1Sd,x2Sd)
> rownames(x) = c(rep(1,15), rep(2,20), rep(3,25))
> plot(x, pch = as.character(rownames(x)), asp = 1)
```

Data with 60 rows and 2 columns are generated with each column from a mixture of three normal distributions are simulated. This type of data can be seen in gene expression or DNA methylation studies on pattern recognitions, e.g., detecting clusters of genes (60 genes in this example with measurements from two subjects) with each cluster sharing comparable expression levels. The scaling done by `x1/sd(x1)` and `x2/sd(x2)`, which removes the effects of different scales in the two variables. To observe the pattern of the original and scaled data, we draw scatter plots using the `plot` function. The argument `asp` is set at 1 to produce plots where distances between points are represented accurately on screen. The clustering pattern in the original data (Figure 5.2a) is not as clear as that from the scaled data (Figure 5.2b).

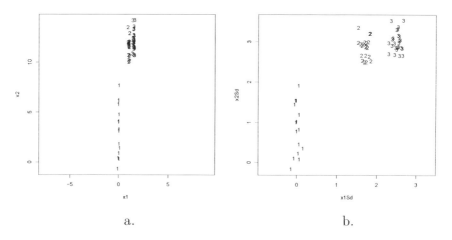

a.   b.

Figure 5.2   Scatter plot of the simulated data, raw (a) and scaled (b) data.

The following codes utilizes the function `kmeans` to cluster scaled data.

```
> x = cbind(x1Sd,x2Sd)
> kmeansClust = kmeans(x, centers = 3, nstart = 2)
> # plot the clusters
> plot(x1Sd, x2Sd, main = 'K-means clustering', cex.main = 2,
+ pch=kmeansClust$cluster)
> legend(2, 1.5, c(1:3), pch = unique(kmeansClust$cluster),
  bty = "n")
```

In the `kmeans` function, the number of clusters is fixed at 3. To determine the number of clusters, the scree test noted earlier can be applied by obtaining `kmeansClust$tot.withinss` for each number of clusters and plotting `kmeansClust$tot.withinss` versus the number of clusters. The default algorithm for clustering is proposed by Hartigan and Wong [64], which in general outperforms the other several methods [102, 41, 105].

It is suggested that we try several random starts by setting `nstart> 1`. In the above, we set `nstart=2`. As seen in Figure 5.3 (generated by the last two lines in the program above), all objects are clustered correctly. We will see later in this section that having data with the same scales will substantially improve the quality of identified clusters.

K−means clustering

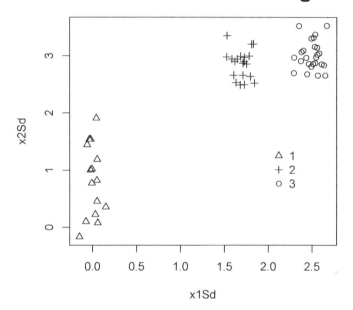

Figure 5.3 Three clusters inferred using the K-means approach.

Another partitioning-based method similar to the K-means is the partition around medoids (PAM) algorithm proposed by Kaufman and Rousseeuw [83]. PAM uses medoids (data points) for centers of each cluster but not mean of data points in a cluster as in K-means. Because of this, PAM is relatively less sensitive to outliers. PAM calculates Silhouette information, S, which can be used to optimize the number of clusters. For each clustering object,

$$
\begin{aligned}
S_i &= 1 - \bar{d}_{\text{within}}/\bar{d}_{\text{between}}, \quad \text{if } \bar{d}_{\text{within}} < \bar{d}_{\text{between}}, \\
&= \bar{d}_{\text{between}}/\bar{d}_{\text{within}} - 1, \quad \text{if } \bar{d}_{\text{within}} > \bar{d}_{\text{between}}, \\
&= 0, \quad \text{if } \bar{d}_{\text{within}} = \bar{d}_{\text{between}},
\end{aligned} \tag{5.1}
$$

where S_i is the Silhouette information for clustering object i and $\bar{d}$ denotes average distance or dissimilarity. Consequently, $\bar{d}_{\text{within}}$ refers to the average distance between clustering object i and other points in the same cluster, and $\bar{d}_{\text{between}}$ is the average distance between clustering object i and the others in other clusters. Silhouette information S_i ranges from -1 to 1, and the higher the S_i the better clustering quality of clustering object i. Averaging S_i across all clustering objects will help us to determine the number of clusters such that within-cluster distances are minimized while maximizing between-cluster distances. Because of the clustering scheme in PAM, PAM can be applied to categorical variables or mixture of continuous and categorical variables with Gower distance utilized. The corresponding R function of PAM is `pam` in the `cluster` package. In this function, the averaged Silhouette information is denoted as "averaged Silhouette width". We use the same data set to demonstrate this function.

```
> library(cluster)
> pamClust = pam(x = cbind(x1Sd,x2Sd), k = 3, diss = FALSE,
+           metric = "euclidean")
> summary(pamClust)[7]$silinfo$avg.width
[1] 0.6466787
```

In the above, we specify the number of clusters as k=3. Since x is a data matrix, the logical flag for x being a dissimilarity matrix is `diss = FALSE`. The dissimilarity is defined as the Euclidean distance in the data. To extract averaged Silhouette width, we use the `summary` statement, `summary(pamClust)[7]$silinfo$avg.width`, based on which we can optimize the number of clusters by setting k at a set of different values and choose k using the scree test noted earlier for K-means.

5.1.3 Hierarchical clustering

Two subcategories of methods are in hierarchical clustering: agglomerative and divisive clustering. Agglomerative clustering starts from singleton clusters, that is, each cluster only includes a singleton. Based on certain criteria, these singleton clusters successively merge (or agglomerate) pairs of clusters until all clusters are merged into one single cluster that contains all the clustering objects. Divisive clustering, on the other hand, starts from one cluster with all clustering objects (either variables or subjects) in the cluster, and the single cluster splits into two clusters following certain criteria defined based on distances. The splitting continues with each cluster split into two clusters until only singleton clusters remain.

Agglomerative clustering In agglomerative clustering, the criteria for merging are determined based on different definitions of distances or dissimilarities between clusters. In the following, we introduce various commonly applied criteria: single linkage, complete linkage, average linkage, Ward's method, and centroid method.

The criterion single linkage focuses on distances between nearest neighbors, and thus methods based on this linkage are also called "nearest neighbor procedure". In particular, the distance between two clusters is defined as minimum distance between every pair of individual clustering objects. For instance, denote by $c_1 = \{x_1, x_2, x_3\}, c_2 = \{x_4, x_5\}$ two clusters with the first cluster containing $\{x_1, x_2, x_3\}$. Following the single linkage criterion, the distance between these two clusters is defined as

$$d_{12} = \min\{d(x_1, x_4), d(x_1, x_5), d(x_2, x_4), d(x_2, x_5), d(x_3, x_4), d(x_3, x_5)\},$$

the minimum distance between every pair of data points in the two clusters. For agglomerative clustering, at each step, combine two clusters that are the most similar. As seen in the definition, the linkage is sensitive to units and scaling. Such a rule of merging is applied in all the hierarchical clustering methods.

Complete linkage measures distance between two clusters from an opposite side. The distance between two clusters is defined as the maximum distance between every pair of individuals. For the two clusters noted above, their distance is defined as $d_{12} = \max\{d(x_1, x_4), d(x_1, x_5), d(x_2, x_4), d(x_2, x_5), d(x_3, x_4), d(x_3, x_5)\}$, which is the distance between the "farthest" neighbors. Similar to the method of single linkage, change of measures or scales may change the the behavior of clustering. Based on the definitions of these two linkages, it is not difficult to see that clustering methods built upon the single linkage criterion has the ability to find non-spherical or non-elliptical (i.e. irregular) clusters, and tends to isolate extreme values into separate clusters. Complete linkage, on the other hand, tends to produce clusters with similar diameters.

Instead of searching for minimum or maximum distances between two clusters as in single and complete linkages, the average linkage criterion focuses on average of distances between clusters. In this criterion, the distance between two clusters is a weighted average distance between individuals in the clusters. Using the same set of two clusters, $c_1 = \{x_1, x_2, x_3\}, c_2 = \{x_4, x_5\}$, the distance between these two clusters is defined as

$$d_{12} = \sum_{i \in c_1} \sum_{j \in c_2} w_{ij} d(x_i, x_j),$$

$$w_{ij} = \frac{1}{n_{c_1} n_{c_2}},$$

where n_{c_1} and n_{c_2} denote the number of clustering objects in each of the two clusters, and thus $n_{c_1} n_{c_2}$ denotes the number of paired objects in total. This criterion tends to produce clusters with similar variances.

Ward's method determines merging based on total variations in the data, evaluated as the sum of squared Euclidean distance from the mean vector (or centroid) of each cluster.

$$SST = \sum_{i=1}^{I} \sum_{j=1}^{n_i} (x_j^i - \bar{x}^i)^T (x_j^i - \bar{x}^i),$$

where I is the number of clusters, n_i is the number of clustering objects in cluster i. As seen from the definition of SST, outliers can strongly impact the quality of clusters.

An approach is less sensitive to outliers is the Centroid method, which evaluates distances between clusters using centroids,

$$d_{ij} = (\bar{x}_i - \bar{x}_j)^T (\bar{x}_i - \bar{x}_j),$$

where $\bar{x}_i$ is the centroid of cluster i, and $\bar{x}_j$ the centroid for cluster j.

Two common R functions are used to perform agglomerative clustering: hclust and agnes. The implementation of function hclust is similar to agnes. In the following, we apply agnes to the simulated data discussed in Section 5.1.2 to demonstrate agglomerative clustering using R. The function agnes is in the cluster package. We first cluster the objects using original data using agnes. The dendrogram of the inferred clusters are in Figure 5.4.

```
> library(cluster)
> x = cbind(x1,x2)
> rownames(x) = c(rep(1,15), rep(2,20), rep(3,25))
> AgnesFit = agnes(x)
> plot(AgnesFit, which.plots = 2, main = '(Unscaled) agnes -
dendrogram', sub = '', cex = 1, cex.main = 2, xlab = '')
```

The default distance metric in `agnes` is Euclidean distance, `metric="euclidean"`. Another option of `metric` is `"manhattan"`. The default criterion for combining two clusters is average linkage, `method="average"`. Other criteria discussed earlier are also available in `agnes`. The `plot` function used here is in the `plot.agnes` package and used to create plots for visualizing the clusters inferred from `agnes`. In this `plot` function, `which.plots` is used to indicate what type of graphs to be drawn with `NULL` to plot both a banner plot and a dendrogram, 1 for a banner, and 2 for a dendrogram. The default of the `sub` argument is to include agglomerative coefficient in the graph. The agglomerative coefficient is used to assess the quality of the clustering. Let $d(i)$ denote the dissimilarity (i.e., distance) of object $i, i = 1, \cdots, n$ to the first cluster it is merged with divided by the dissimilarity of the merger in the last step of the algorithm. The agglomerative coefficient is defined as the average of all $1 - d(i)$. The range of the agglomerative coefficient is between 0 and 1, the higher the better. The statistics agglomerative coefficient is only useful to compare same-size data sets (sensitive to sample size) and seems not sensitive enough to mis-classification. For this reason, Figure 5.4 does not include this statistics. With the original scale, apparently several clustering objects are misplaced due to the sensitivity of the criterion used to split (Figure 5.4).

We next apply `agnes` to scaled data, x1Sd and x2Sd,

```
> x1Sd = x1/sd(x1)
> x2Sd = x2/sd(x2)
> xSd = cbind(x1Sd,x2Sd)
> rownames(xSd) = c(rep(1,15), rep(2,20), rep(3,25))
> AgnesFit1 = agnes(xSd)
> plot(AgnesFit1, which.plots = 2, main = '(Scaled) agnes -
dendrogram', sub = '', cex = 1, cex.main = 2, xlab = '')
```

and after scaling, the objects are all clustered correctly (Figure 5.5). Almost all clustering methods are sensitive to measures or scales, and it is thus recommended to scale the data or standardize the data before performing cluster analyses, especially when the clustering objects are not with the same units.

Divisive clustering For divisive clustering, the five criteria discussed earlier are applicable when splitting a cluster except for the complete linkage. In this case, the splitting criterion is to the smallest distance of two candidate subsets, instead of the largest between-cluster distance.

(Unscaled) agnes − dendrogram

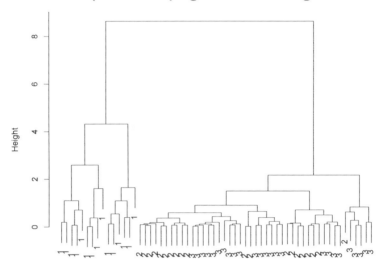

Figure 5.4   Dendrograms of clusters inferred by agnes using original data.

The commonly used divisive clustering approach was the DIANA algorithm, presented in Chapter 6 of Kaufman and Rousseeuw [84] and available in R, the function diana. To choose a cluster to split at each stage, the largest distances (or dissmilarity) between any two clustering objects within a cluster is used and such a largest distance is named as diameter. The cluster with the largest diameter is selected to split. To divide the selected cluster, the method first identifies a clustering object such that the average of its distances to the other clustering objects in the cluster is the largest. This clustering object initiates a "splinter group". This is then followed by selecting clustering objects from the cluster ("parent cluster") such that they are closer to the "splinter group" than to the "parent group". This process continues until no points can be moved to the "splinter group" and the cluster is now split into two clusters.

We again use the simulated data x1 and x2 to demonstrate the use of diana. As noted earlier, scaled data are recommended in cluster analyses, and thus in the following diana is applied to x1Sd and x2Sd.

```
> xSd = cbind(x1Sd,x2Sd)
> DianaFit = diana(xSd)
> plot(DianaFit,which.plots = 2,
+     main = "(Scaled) diana - dendrogram", sub = '',
```

(Scaled) agnes – dendrogram

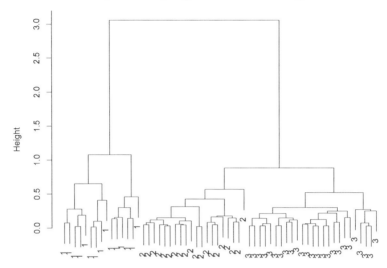

Figure 5.5 Dendrograms of clusters inferred by agnes using scaled data.

```
+       cex = .8, cex.main = 2,xlab = '')
```

The arguments included in diana overall are the same as those in agnes such as distance metrics, except that diana does not provide an option on setting splitting criteria. As seen in the results from agnes, with scaled data, the method in diana also clusters all the objects into the right cluster (Figure 5.6).

5.1.4   Hybrids of partitioning-based and hierarchical clustering

Approaches have been developed that are a mixture of partitioning-based and hierarchical clustering, or a mixture of agglomerative and divisive clustering. For approaches that combine partitioning and hierarchical clustering, they in general start the root node with all clustering objects included, select the best partition of the objects, followed by the possibility of collapsing one or more pairs of clusters. We introduce one type of such approaches, the hierarchical ordered partitioning and collapsing hybrid (HOPACH) method [142]. It is a mixture of divisive, partitioning, and agglomerative clustering. For approaches jointly using agglomerative and divisive clustering, they are usually built upon divisive clustering and start from a single cluster, split a cluster into two but

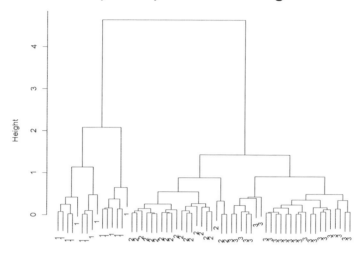

Figure 5.6 Dendrograms of clusters inferred by `diana` using scaled data.

allow subclusters or children clusters to merge or do not split clusters if they should not be split. For methods in this category, we focus on the hybrid hierarchical clustering via mutual clusters (hybridHclust in R) approach, where a mutual cluster is a set of clustering objects closer to each other than to all other objects, and thus a cluster that should never be broken [20].

The HOPACH method   This clustering algorithm is a hybrid between an agglomerative and a divisive algorithm, and with partitioning algorithm built in to determine the best split at each step. It starts from a single cluster with all clustering objects (root node) and, at each level, collapsing steps are applied to unite two similar clusters. For the partitioning algorithm, the partition around medoids (PAM) algorithm proposed by Kaufman and Rousseeuw [83] is applied.

The median (or mean) split silhouette (MSS) criterion improved from the Silhouette information in PAM is proposed in HOPACH to determine the number of subclusters, optimize partitions and select pairs of clusters to collapse at each level. The goal is to minimize MSS, which is a measure of cluster heterogeneity [142], and find small clusters if any inside big clusters. For each clustering object, the HOPACH algorithm measures how well matched it is to the other objects in the cluster compared to if it were switched to the next closest cluster. This is evaluated using the Silhouette information defined in (5.1). Using the average of the Silhouettes to determine the quality of cluster and optimize the splitting of a cluster to subclusters by maximizing mean Silhouettes. After splitting, each clustering object's Silhouette information is re-evaluated relative to their sibling clusters, and a median of Silhouettes, named as median split silhouette (MSS) [142], is taken within each parent cluster. The median of Silhouettes across all parent clusters is an overall measurement for homogeneity within parent clusters. Two parent clusters can be collapsed if doing so will improve MSS.

An R package, hopach, which implements this algorithm is available in Bioconductor. In the following, we apply the package to cluster the simulated data discussed above.

```
> library(hopach)
> hopClust = hopach(xSd, d = 'euclid')
> # Look at num. of clusters
> hopClust$clust$k
[1] 24
> # Look at sizes of clusters
> table(hopClust$clust$sizes)

 1  2  3  4  9
 7  5 10  1  1
```

In the above, the clustering is applied to the rows of scaled data xSd. The Euclidean distance metric, indicated by d='euclid' in the function hopach, is used to determine the distance between clustering points. We can use hopClust$clust$k to see the number of clusters and table(hopClust$clust$sizes) to check the size of each cluster.

By use of this hybrid clustering method, 24 clusters are identified, including 7 clusters with singletons. The large number of clusters is expected since this clustering method is built upon divisive clustering. To visualize the clusters, heat maps can be used, which is helpful to tease out overall clustering patterns instead of the more scrutinized 24 clusters. The R function dplot can be used to for this purpose. A heat map is generated based on ordered dissimilarity matrix. In dplot, two ways are available to define the order of the dissimilarity matrix, one being the order of elements in the final level of the hierarchical tree, and the other being the order from the level of the tree corresponding to the main clusters. To apply dplot to the above hopClust object,

```
> library(RColorBrewer)
> source("https://bioconductor.org/biocLite.R")
> biocLite("bioDist")
> library(bioDist)
> colRB = colorRampPalette(brewer.pal(10, "RdBu"))(256)
> rownames(xSd) = seq(1,nrow(xSd))
> distanceEuc = distancematrix(xSd, d = "euclid")
> dplot(distanceEuc, hopClust,lab = rownames(xSd),
ord = 'final', showclust = FALSE,
+ col = colRB,main = 'HOPACH - Euclidean distance')
```

Instead of the default red-yellow color theme supplied by dplot, a red-blue color scheme defined by colorRampPalette in the RColorBrewer package is used in the plot. A color of red represents high similarity and a color blue refers to high dissimilarity. A distance matrix needs to be calculated and included in the dplot function. To be consistent with the distance metric used to cluster xSd, an Euclidean distance matrix is calculated using the function distancematrix in the hopach package. In the above, the elements in the dissimilarity matrix is ordered based on the final levels in the tree, ord="final". Running the dplot function, the ordered dissimilarity matrix is shown as a heat map and clusters appear as blocks on the diagonal of the heat map. For the simulated data, a clear pattern of 3 clusters is shown (Figure 5.7), Note that the color is not shown in the figure due to the black-white printing style.

The hybrid hierarchical clustering method This divisive clustering method has a feature of maintaining a group of clustering objects denoted as mutual clusters [20]. To define a mutual cluster, let S denote such a cluster. For any object x in S, we have $d(x,y) > \max_{w \in S, z \in S} d(w, z)$ for any object y outside S. This definition claims a subset of clustering objects being a mutual cluster if the minimum distance between any object in the mutual cluster and an object outside the mutual cluster is larger than the maximum distance between points in the mutual cluster. These mutual clusters cannot be further divided, but within each mutual cluster, clustering is performed to create hybrid clusters. The algorithm has three steps:

1. Identify mutual clusters using the distance-based definition of such clusters.

2. Perform a constrained divisive clustering such that each mutual cluster must stay intact. The divisive clustering is achieved by use

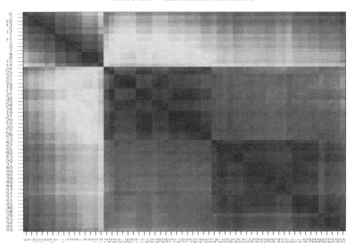

Figure 5.7   Heat map of ordered dissimilarity matrix showing 3 clusters.

of the K-means approach at each node, which is noted as "tree-structured vector quantization" (TSVQ) in the area of engineering [54].

3. After divisive clustering is complete, within each mutual cluster, hybrid clusters are inferred. Since mutual clusters are often small, similar results will be obtained from divisive or agglomerative methods. However, since divisive clustering is implemented in step 2, it is simpler to use the same algorithm when determining the clusters within each mutual cluster.

This algorithm is implemented in R with the hybridHclust package, which detects the mutual clusters and divisively clusters points with mutual clusters protected. We continue to use the simulated data x1 and x2 to demonstrate the implementation of this package with scaled data, x1Sd and x2Sd.

```
> library(hybridHclust)
> x = cbind(x1Sd,x2Sd)
> rownames(x) = c(rep(1,15), rep(2,20), rep(3,25))
> hyb = hybridHclust(x)
```

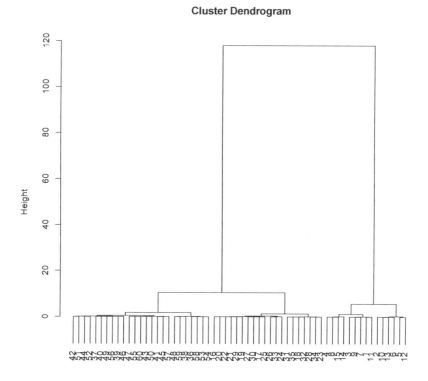

Figure 5.8 Dendrogram of the clusters detected by use of mutual clusters.

In the above program, `hyb` is a dendrogram and we can use `plot(hyb,rownames(x))` to examine the quality of clustering.

The function `mutualcluster` outputs the identified mutual clusters:

```
> mc1 <- mutualCluster(x)
> mc1
1 :  16 17 18 19 20 21 22 23 24 25 26 27 28 29 30 31 32 33 34 35
     36 37 38 39 40 41 42 43 44 45 46 47 48 49 50 51 52 53 54 55
     56 57 58 59 60
2 : 5 12
3 : 2 10 13
4 : 1 11
5 : 3 9
6 : 4 8 15
```

Note that observations 16 to 60 are included in one mutual cluster. As seen in the cluster dendrogram plot (Figure 5.8), these 45 observations are further split into two clusters. The inclusion of these observations in one mutual cluster implies shorter distances between these observations, compared to observations outside this mutual cluster. This is supported by the plot of data shown in Figure 5.2a.

5.1.5 Examples – clustering to detect gene expression patterns

In this section, we apply the clustering methods discussed so far to a data set of gene expressions from the RNA-Seq technique. The data are from the study in Brooks et al. [13]. The authors combined RNAi and RNA-Seq to identify exons regulated by Pasilla, the *Drosophila melanogaster* ortholog of mammalian NOVA1 and NOVA2. The RNA-Seq data and related information are available from the Gene Expression Omnibus (GEO) database under accession numbers GSM461176-GSM461181. The data are downloaded from https://figshare.com/s/e08e71c42f118dbe8be6. We first read in the data and perform necessary quality controls. The reads were aligned to the Drosophila reference genome and summarized at the gene level.

File counts_Drosophila.txt contains reads in counts of $14,869$ genes on seven samples. Information on sample IDs, treatment status, and depth in sequencing indicated by Library are saved in SampleInfo_Drosophila.txt. The abbreviation PE denotes paired end sequencing and SE is for single end sequencing.

```
> # read in the RNA-Seq data
> Counts = read.delim(file="counts\_Drosophila.txt")
> # Information on the data including
> # treatment status and library (sequencing depth)
> Targets = read.delim(file = "SampleInfo_Drosophila.txt")
> dim(Counts)
[1] 14869     7
> Targets
  SampleName      Group Library
1  SRR031714 Untreated      PE
2  SRR031716 Untreated      PE
3  SRR031724   Treated      PE
4  SRR031726   Treated      PE
5  SRR031708 Untreated      SE
6  SRR031718   Treated      SE
7  SRR031728 Untreated      SE
```

To filter out genes with low counts, we compute the number of reads per million reads mapped and calculated by the library size of each sample, using the R function cpm in the edgeR package (it also requires the limma package). This is then followed by specifying a threshold of counts per million (cpm) for low reading, which is selected corresponding to $10-15$ reads. In the following, the threshold is set at 3, corresponding to 12 reads (Figure 5.9).

```
> library(limma)
> library(edgeR)
> # compute counts per million reads mapped
> CountsRate = cpm(Counts)
> plot(Counts[,1],CountsRate[,1],xlim = c(0,20),ylim=c(0,5))
> abline(v = 12,col = 2)
> abline(h = 3,col = 4)
> # 3 counts per million is used as the cutoff
> # corresponding to 12 absolute reads
> Cutoff = CountsRate > 3
```

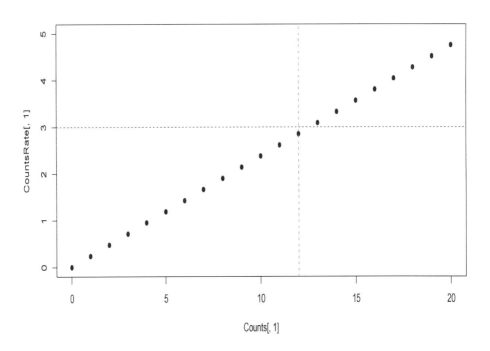

Figure 5.9 Plot of counts per million vs. number of reads. The vertical dashed line corresponds to 12 reads.

After the decision of threshold, we exclude genes such that 3 cpm observed in at least three samples. This is achieved by use of function rowSums. The screening step excluded more than half of the genes and in total 7245 genes are included in the subsequent analyses.

```
> Screen = rowSums(Cutoff) >= 3
> table(Screen)
Screen
FALSE   TRUE
 7624   7245
>
> CountsKeep = Counts[Screen,]
> dim(CountsKeep)
[1] 7245    7
```

To adjust the impact of library size, we normalize the counts by the sample normalization factor available in the edgeR package.

```
> y = DGEList(CountsKeep)
> # calculate log base 2 of the counts with counts normalized
> # by the sample normalization factor.
> NormLogCounts = cpm(y,log = TRUE)
```

For the purpose of demonstration, we choose the top 50 genes such that they show the largest variations across the 7 samples. The variances are calculated using function apply.

```
> VarGenes = apply(NormLogCounts, 1, var)
> # get the top 50 gene names
> SelectVar = names(sort(VarGenes, decreasing = TRUE))[1:50]
> HighVarCountsRate = NormLogCounts[SelectVar,]
> dim(HighVarCountsRate)
[1] 50   7
```

We now have normalized cpm of 50 genes in 7 samples. The cluster analyses will focus on the genes and we start from the K-means approach. As done earlier, we create a scree plot on the total within cluster sum of squares to assist the decision on the number of clusters. As shown in Figure 5.10, the turning point of within cluster variations is at the number of 4 clusters.

```
> max<-10
> ssWithin = rep(0,(max-1))
> for (i in 2:max)
```

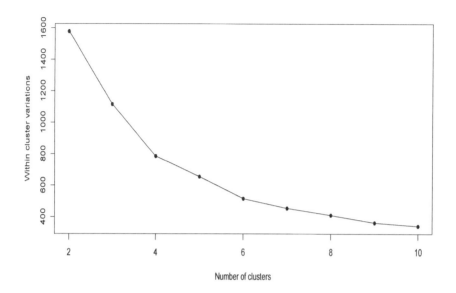

Figure 5.10 Scree plot of the within cluster sum of squares (K-means).

```
+ {
+ set.seed(i*5000)
+ kClust = kmeans(x = HighVarCountsRate,centers = i)
+
+ ssWithin[i-1] = kClust$tot.withinss
+ }
> # choose the number of clusters minimizing within cluster
variations
> plot(seq(2,max), ssWithin, pch = 19, cex = 0.8,
xlab = "Number of clusters", + ylab = "Within cluster
variations")
> lines(seq(2,max), ssWithin)
```

We then apply kmeans again to the RNA-Seq data of 50 genes with centers=4. The clusplot in the R package cluster performs principal components analysis and displays the clusters based on the first two components which explain 76.22% about the variation in the data. The data are overall well grouped using four clusters (Figure 5.11). The shades with different density are achieved by setting shade=TRUE. A unit area with more data points is associated with a higher density.

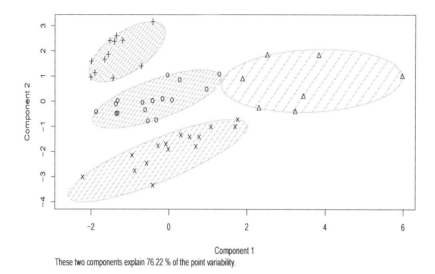

Component 1
These two components explain 76.22 % of the point variability.

Figure 5.11 Display of the four clusters using the first two principal components (K-means).

```
> kmeansClust = kmeans(x=HighVarCountsRate,centers = 4,
nstart = 2)
> library(cluster)
> clusplot(HighVarCountsRate,kmeansClust$cluster,shade = TRUE,
+ color = TRUE,col.clus = rep("gray",4), col.p = "black",
lines = 0,main = NULL)
```

To use PAM to cluster the data, the program is similar to what has been introduced earlier using simulated data. To choose the best number of clusters, we utilize grid search to maximize average Silhouette width (Figure 5.12). Using PAM, 5 clusters are identified (Figure 5.13), but the separation of the data is not as good as the result based on K-means.

```
> # using PAM to cluster
> library(cluster)
> SiWidth = rep(0,(max-1))
> for (i in 2:max)
+ {
+ pamClust = pam(x=HighVarCountsRate, k = i, diss = FALSE,
+          metric = "euclidean")
+ SiWidth[i-1] = pamClust$silinfo$avg.width
+ }
```

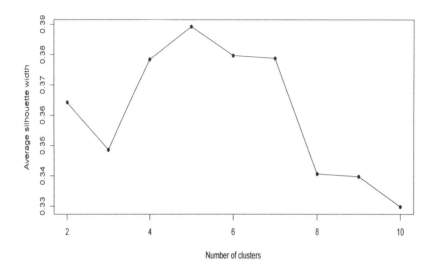

Figure 5.12 The scree plot based on average silhouette width (PAM).

```
> # Choose the number of cluster maximizing
> # the average silhouette widths
> plot(seq(2,max),SiWidth, pch = 19, cex = 0.8, xlab = "Number of clusters",
+ ylab = "Average silhouette width")
> lines(seq(2,max),SiWidth)

> pamClust = pam(x=HighVarCountsRate, k = 5, diss = FALSE,
+            metric = "euclidean")
> dev.new()
> clusplot(HighVarCountsRate,pamClust$cluster,shade=TRUE,
+ color=TRUE,col.clus=rep("gray",4),col.p="black",
lines=0,main=NULL)
```

For clusters analyzed using agnes and diana, for the purpose of comparison with results from partitioning-based methods, we first estimate the best number of clusters using the NbClust function in the package NbClust. In the following codes, the best number of clusters is estimated as 6, using Silhouette index with clusters inferred based on Euclidean distance using Ward's method.

```
> library(NbClust)
> hierarClust = NbClust(data = HighVarCountsRate,
```

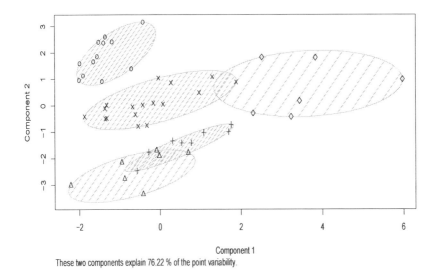

Figure 5.13 Display of the five clusters using the first two principal components (PAM).

```
+ distance = "euclidean",
+ min.nc = 2, max.nc = 10, method = "ward.D",
+ index = "silhouette")
> hierarClust$Best.nc
Number_clusters     Value_Index
        6.0000          0.3778
```

We next trim the tree inferred using Ward's method in agnes. The function cutree is applied to extract a sub-tree. Function clusplot is not applicable to hierarchical clusters. To visualize the clusters, we use the fvix_cluster function in the factoextra package (it also needs the ggplot2 package). This function uses the top two principal components to display the group patterns. The loading of each gene is plotted in points specified by geom="point" and each cluster is framed with different shapes. Frame types are specified by ellipse.type. In each cluster, a point with a larger size represents the medoid of that cluster. The detected six clusters by agnes have some overlaps and it maybe reasonable to collapse some clusters (Figure 5.14).

```
> # hierarchical clustering using agglomerative nesting (agnes)
```

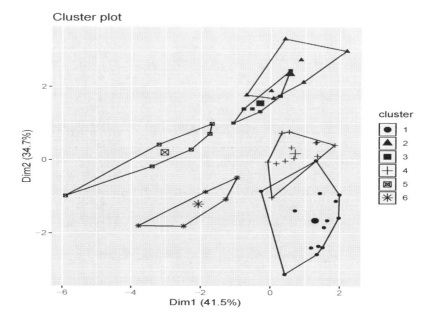

Figure 5.14 The six clusters identified using agnes.

```
> AgnesFit = agnes(HighVarCountsRate,method="ward")
> subTree = cutree(as.hclust(AgnesFit), k = 6)
>
> # plot the clusters
> library(factoextra)
> library(ggsignif)
> fviz_cluster(list(data = HighVarCountsRate, cluster = subTree),
+ geom="point", ellipse.type = "convex",
+ palette = rep("black",6), ggtheme = theme_gray())
```

For divisive cluster analysis, we essentially follow the same procedure. The function diana identified a singleton cluster and genes in other clusters are better separately compared to the results from agnes (Figure 5.15).

Next we detect clusters of genes considering the existence of mutual clusters. The coding is similar to what is discussed earlier. Here we focus on the detected mutual clusters as well as the dendrogram of the results. In total, 14 mutual clusters are detected. To plot the dendrogram with color indicating different clusters, we use ggplot. We need to first convert the object from hybridHclust to a dendrogram using the as.dendrogram function. Next we specify the number of clusters k

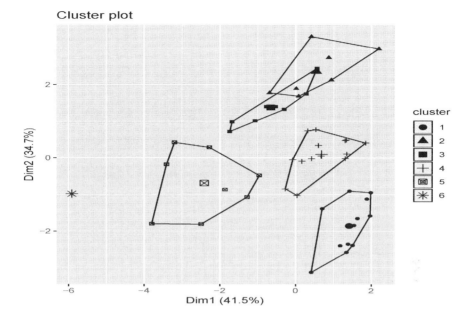

Figure 5.15 The six clusters identified using `diana`.

and colors used to mark clusters. Finally, the function `ggplot` is utilized to plot the dendrogram (Figure 5.16). Based on the dendrogram, it is reasonable to conclude that 6 clusters can reasonably separate the 50 genes.

```
> mc1 <- mutualCluster(HighVarCountsRate)
> mc1
1 : 2 4
2 : 3 5
3 : 6 7 8
4 : 11 15
5 : 29 34
6 : 12 22
7 : 20 49
8 : 45 48
9 : 32 43
10 : 23 28
11 : 18 19
12 : 37 41
13 : 27 39 42 50
14 : 30 36 46
> hyb1 <- hybridHclust(HighVarCountsRate,mc1)
```

```
> library("reshape2")
> library("purrr")
> library("dplyr")
> library("dendextend")
> dendro = as.dendrogram(hyb1)
> dendro.col <- dendro %>%
+    set("branches_k_color", k = 6, value =
+    rep(c("black","gray"),3)) %>%
+    set("branches_lwd", 0.8) %>%
+    set("labels_colors",
+        value = c("darkslategray")) %>%
+    set("labels_cex", 1)
> ggd1 <- as.ggdend(dendro.col)
> ggplot(ggd1, theme = theme_minimal()) +
+    labs(x = "Gene numbers", y = "Height", title = "6 clusters")
Warning message:
Removed 99 rows containing missing values (geom_point).
```

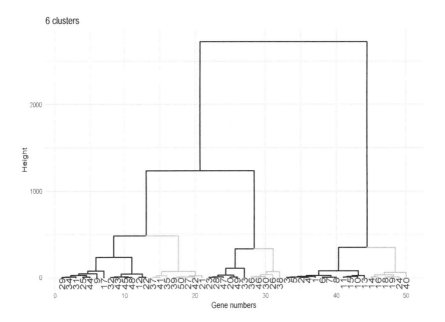

Figure 5.16 The six clusters identified using `hybridHclust` with mutual clusters.

5.2 CLUSTER ANALYSES IN LINEAR REGRESSIONS

In general, cluster analyses in linear regressions are designed to group variables, instead of subjects, such that each cluster of variables has consistent associations with one or more external variables of interest. As a supervised machine learning technique, this type of clustering is commonly applied to identify genes or other genetic features. We focus on two approaches, one is by Qin and Self [115], and the other by Yengo et al. [151].

Let $\boldsymbol{y}$ denote a matrix of data on K variables to be clustered each with measures on n subjects. The data matrix $\boldsymbol{y}$ is of dimension $n \times K$. In addition, we have a matrix of data on m covariates for each subject, denoted by $\boldsymbol{x}$ with dimension of $n \times m$. The goal is to put the K variables into C clusters based on their associations with $\boldsymbol{x}$ such that variables in the same cluster share the same values of the regression coefficients. We start introducing the method from linear regressions, that is,

$$
\begin{aligned}
y_{ik}|(\boldsymbol{\mu}_k &= c, \boldsymbol{x}_{ik}) = \boldsymbol{x}_{ik}^T \boldsymbol{\beta}_c + \epsilon_{ik} \\
\epsilon_{ik} &\sim N(0, \sigma_c^2),
\end{aligned}
$$

where $\boldsymbol{\mu}_k = (\mu_{k,1}, ..., \mu_{k,C})$ is a vector of indicators denoting which cluster variable k belongs to, and σ_c^2 is the variance of random error ϵ for cluster c. Qin and Self [115] proposed an expectation-maximization (EM) algorithm to infer the parameters including cluster assignment. To determine the number of clusters, Bayesian information criterion (BIC) can be applied for the purpose of goodness of fit. Qin and Self [115] proposed an accompanying criterion, bootstrapped maximum volume (BMV), measuring stability of cluster centers for each given C, $BMV_C = max\{volume(\hat{\Sigma}_c), c = 1, \cdots, C\}$, where $\hat{\Sigma}_c$ is an estimated covariance matrix of the inferred regression coefficients for each of the C clusters. Bootstrap samples can be used to estimate $\hat{\Sigma}_c$. A volume is defined as the largest eigenvalue of $\hat{\Sigma}_c$. The number of clusters is selected based on a good fit to the data and stable cluster centers, that is, clusters with a large BIC and small BMV.

This method is programmed into an R function `cluster.reg` in the `RegClust` package. We demonstrate the package via simulated data, which includes 23 variables to be clustered and 2 continuous covariates. This example is adapted from the example provided by the `RegClust` package.

The simulated data has two clusters with coefficients specified by `beta0` as the common intercept, coefficient `beta1` for cluster 1 and coefficient `beta2` for cluster 2. Sample size is set at n=200.

```
> library(RegClust)
> set.seed(1234)
> beta0 = 1
> beta1 = c(0.02, 0.1)
> beta2 = c(-0.04, 0.1)
> n = 200
```

The following lines simulate the two clusters with first cluster having 8 variables and the second 15 variables. The variances of the random errors are unique to each cluster.

```
> # Generate covariates.
> cov1 = runif(n, min = 18, max = 70)
> cov2 = rnorm(n,-3,3)
> cov = cbind(cov1, cov2)
>
> # The first cluster, which has nyc1 variables
> nyc1 = 8
> yc1 = matrix(NA, n, nyc1)
> for (k in 1:nyc1)
+ {
+   set.seed(1234)
+   yc1[,k] = beta0+cov%*%beta1
+   set.seed(k*100)
+   yc1[,k] = yc1[,k]+rnorm(n)
+ }
>
> # The second cluster, which has nyc2 variables
> nyc2 = 15
> yc2 = matrix(NA, n, nyc2)
> for (k in 1:nyc2)
+ {
+   set.seed(1234)
+   yc2[,k] = beta0+cov%*%beta2
+   set.seed(k*200)
+   yc2[,k] = yc2[,k]+rnorm(n,0,0.5^0.5)
+ }
>
> X = as.matrix(cov)
> Y = as.matrix(cbind(yc1, yc2))
> colnames(Y) = paste(rep("y", ncol(Y)), seq(1,ncol(Y)), sep="")
```

To identify the clusters of Y based on their associations with X, we use cluster.reg.

```
> run = cluster.reg(Y,X)
```

```
Converged at iteration  6 , BIC = -2314.046
> run
$cluster
      y1 y2 y3 y4 y5 y6 y7 y8 y9 y10 y11 y12 y13 y14 y15
[1,]   2  2  2  2  2  2  2  2  1   1   1   1   1   1   1
      y16 y17 y18 y19 y20 y21 y22 y23
[1,]   1   1   1   1   1   1 1   1


$param
          Intercept         cov1       cov2      sigma         pi
cluster 1  1.022655 -0.04032938 0.1053631 0.4762371 0.6521739
cluster 2  1.067028  0.01951102 0.1117932 0.9389717 0.3478261


$likelihood
[1] -1151.725


$BIC
[1] -2314.046
```

The object after running `cluster.reg` includes cluster assignment (`run$cluster`), parameter estimates (`run$param`) including estimates of random error variances and probabilities of cluster assignments, log-likelihood, and the optimized BIC.

The `RegClust` package only handles linear regressions. An R function `fit.CLMM` in the `CORM` package has the ability to perform cluster analyses on variables with repeated measures using linear mixed models [127, 115]. The method in the `fit.CLMM` package allows linear or non-linear associations between y and covariates x with random effects of z [127, 115],

$$
\begin{aligned}
y_{ik}|(\mu_k &= c, x_{ik}) = f(x_{ik}^T, \beta_c) + z_{ik}^T u_{ik} + \epsilon_{ik} \\
u_{ik} &\sim MVN(0, D_c), \\
\epsilon_{ik} &\sim N(0, \sigma_c^2),
\end{aligned}
$$

where D_c is the covariance matrix for cluster c for random effects of u_{ik} and $f(x_{ik}, \beta_c)$ is a function of $x_{ik}, \beta_c)$, e.g., splines. In genetic and epigenetic studies, this type of clustering particularly benefits studies focusing on genetic variants over time such as longitudinal gene expression levels in reaction to a specific treat, in which case the changes of expression levels are likely to be non-linear over time. The R function `fit.CLMM` performs cluster analyses on variables with repeated measures.

The fixed effects and random effects in the linear mixed effects model implemented in fit.CLMM are both set to be the spline basis of time.

We use the example given in [127, 114] to demonstrate the implementation of function fit.CLMM. The codes are adapted from the program in the manual of the CORM package. The data is from a yeast cell cycle regulation study [163]. In total, the first 16 time points of expressions measured at 7-minutes intervals of 256 genes whose transcription is cell cycle dependent are included in the cluster analysis. Patterns of gene expressions over time are used to group the genes.

```
> library(CORM)
> #test data
> data(YeastCellCycle)
> data.y = YeastCellCycle$normalizedData
> data.x = YeastCellCycle$designMatrix
```

The CORM library also includes a subset of expression data which includes 64 genes instead of 256, YeastCellCycle$normalizedData.sample. To test out the function fit.CLMM, the smaller data set will be handy.

```
> #fit the model with different number of clusters and record
+ the log-likelihood
> lh = NULL
> start = 2
> end = 12
> for (i in start:end)
> {
>   n.clst = i
>   fit1 = fit.CLMM(data.y, data.x, data.x, n.clst)
>   lh = c(lh,fit1$theta.hat$llh)
> }
> # plot the log-likelihoods v.s. number of clusters
> linearModel = lm(lh~seq(start, end))
> plot(seq(start,end), lh, type = "n")
> points(seq(start, end), lh)
> lines(seq(start, end), (linearModel[1]$coeff[1]+seq(start,end)*
> + linearModel[1]$coeff[2]))
```

To empirically select the best number of clusters, we can apply the scree plot technique on the log-likelihoods to determine the best number of clusters. In the above, log-likelihood is calculated for the number of clusters varying from 2 to 12. In Figure 5.17, using a linear regression line to assist the selection of the best number of clusters based on the

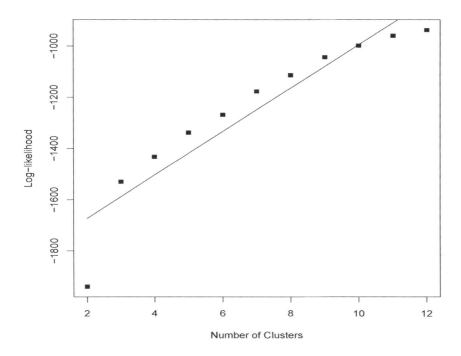

Figure 5.17 Scree plot on the log-likelihoods vs. number of clusters.

elbow rule, we conclude 6 or 7 clusters that may fit the best to the data. After plotting the clusters, it seems 7 clusters are a better choice, as they differentiate the patterns more clearly and the size of each cluster is still reasonably large. Qin et al. [114] suggested 8 clusters, determined based on data separation as well as biological interpretation.

In the following, we include the R codes used to produce plots of patterns of 7 clusters overlaid on plots of the data (Figure 4.18).

```
> n.clst = 7
> fit1 = fit.CLMM(data.y, data.x, data.x, n.clst)
> # extract cluster memberships and put them in order
> fit1.u = apply(fit1$u.hat, MARGIN = 1, FUN = order,
+ decreasing = TRUE)[1,]
> # extract the estimated fixed effects
> zeta.fitted = fit1$theta.hat$zeta.hat
> # calculate the profile for each cluster
> profile = data.x[1,,] %*% t(zeta.fitted)
> #display the profile of each cluster
```

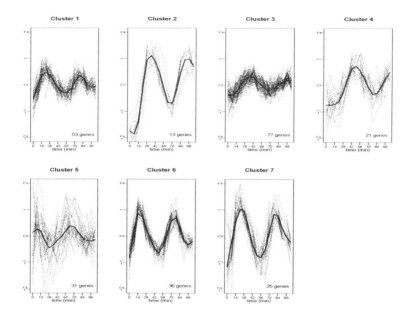

Figure 5.18 Plot of cluster profiles of 7 clusters of 256 genes based on gene expressions.

```
> n.knots = 7
> plot.x = n.knots*(1:dim(data.y)[3]-1)
> par(mfrow = c(2, ceiling((n.clst)/2)),mai = c(0.5,0.5,0.5,0.1),
+ mgp = c(1,0.3,0))
> for(k in 1:n.clst){
+ # plot the fitted cluster-specific profiles
+ plot(plot.x,profile[,k],type = "l", ylim = c(-2,2), main =
+ paste("Cluster",k),
+ xlab = "time (min)", ylab = NA, xaxt = "n", lwd = 2)
+ axis(side = 1, at = plot.x[(1:8)*2-1], labels = paste(plot.x[(1:8)*2-1]),
+ cex.axis = 0.8)
+ # plot the observed profiles for genes in this cluster
+ i.plot = (1:dim(data.y)[1])[fit1.u==k]
+ for(j in i.plot) {
+ lines(plot.x, data.y[j,1,], lty = 3, lwd = 1)}
+ text(84,-1.9, paste(length(which(fit1.u==k)),"genes"))
+ }
```

We must note that the above two clustering techniques can also be used to cluster subjects if the data matrix is $t(n \times 1)$. In other words, for data without repeated measures and the goal is to cluster subjects, we can then transpose our data and apply these two functions, `cluster.reg` and `fit.CLMM` to infer clusters of subjects.

5.3 BICLUSTER ANALYSES

For the bicluster analyses, we focus on two methods implemented in the `biclust` package, the Bimax biclustering (`BCBimax`) and the Plaid Model biclustering (`BCPlaid`) methods [113, 89]. The Bimax biclustering approach applies to binary data, e.g., for a certain number of genes, change and no change in gene expressions compared to subjects in controls.

Bimax biclustering Considering the data to be clustered as a matrix composed of 1's and 0's, E, the Bimax algorithm is to partition the matrix into submatrices starting from diving E into two submatrices by taking one row as a "template". This is then followed by sorting the rows based on the order of fitness to the "template", leading to two better organized submatrices. The algorithm is then recursively applied to each of the two submatrices to further detect biclusters. Within each submatrix, matrices with only zeros will be intact in subsequent clustering. The recursion ends if the current matrix only has 1's. If submatrices do not have any overlaps, the recursive biclustering can be conducted within each submatrix. If submatrices share some rows, special care is needed to ensure inclusion-maximal, that is, any biclusters are not entirely contained in any other biclusters.

We use a subset of the same 16 time points of normalized expressions data of 256 genes such that data of the first 9 genes at 10 time points are included in the biclustering process. The goal is to group genes such that their expressions agree with each other across a set of consecutive time points in terms of direction of expression compared to the controls. To fit the data format required for Bimax biclustering, we dichotomize the data such that positive values are coded as 1 and negative values as 0, using an R function `binarize`.

```
> matrixY = data.y[,1,]
> BinaryGE = binarize(matrixY,0)
> partBinaryGE = BinaryGE[1:9,1:10]
> partBinaryGE
      [,1] [,2] [,3] [,4] [,5] [,6] [,7] [,8] [,9] [,10]
[1,]    0    1    1    1    1    0    0    0    0     0
[2,]    0    0    0    0    1    1    1    1    0     0
[3,]    0    0    0    0    1    1    1    1    0     0
[4,]    0    0    0    1    1    1    1    0    0     1
[5,]    0    0    1    1    1    0    0    0    0     1
[6,]    1    0    0    0    0    1    1    1    1     1
```

```
[7,]    0    0    0    0    1    1    1    1    1    0
[8,]    0    0    0    0    1    1    1    1    0    0
[9,]    0    0    0    1    1    1    1    1    0    0
```

In the above, `partBinaryGE` is the dichotomized data by using 0 as the cutoff applied to the standardized expressions of 9 genes across 10 time points. After this step, the function `biclust` is then implemented to perform bicluster analyses on the dichotomized data.

```
> res = biclust(x = partBinaryGE, method = BCBimax(),
+ minr = 2, minc = 3, number = 20)
> res@Number
[1] 8
```

We choose the method by setting `method=BCMimax()`. Options `minr` and `minc` indicate the minimum numbers of rows and columns for each bicluster. These two parameters are pre-specified based on patterns of data and our preference on size of clusters. After observing the data, to have more detailed clustering, we set `minr=2` and `minc=3`. The setting `number` indicate the maximum number of clusters allowed. For the data above, the Bimax biclustering identified 8 biclusters.

To observe bicluster assignment, we use two slots available in the Biclust object `res`, `res@RowxNumber` which gives the cluster assignment of the 9 rows (9 genes) and `res@NumberxCol` which gives the bicluster assignment of the 10 columns of the data (10 time points).

```
> # labels of cluster assignment of rows (9 rows), for
+ instance, "3" means in cluster 3.
> rowMatrix = matrix(rep(seq(1,res@Number), nrow(partBinaryGE)),
+ ncol = res@Number, byrow=TRUE)
> (res@RowxNumber*1)*rowMatrix
      [,1] [,2] [,3] [,4] [,5] [,6] [,7] [,8]
 [1,]    1    0    0    0    0    0    0    0
 [2,]    0    2    0    0    0    6    7    0
 [3,]    0    2    0    0    0    6    7    0
 [4,]    0    0    3    0    5    6    0    8
 [5,]    1    0    0    0    5    0    0    0
 [6,]    0    2    3    4    0    0    0    0
 [7,]    0    2    0    4    0    6    7    0
 [8,]    0    2    0    0    0    6    7    0
 [9,]    0    2    0    0    0    6    7    8
```

From the results above, it is clear that one gene can be in multiple biclusters, which is understandable since its expressions may show different patterns at different time frames, for instance, gene 9 is in biclusters 2, 6, 7, and 8. However, further assessment on the bicluster assignment on the times may reflect some ambiguity generated by this Bimax biclustering algorithm.

```
> # labels of cluster assignment of columns (10 columns), for
+ instance, "3" means in cluster 3.
> colMatrix = matrix(rep(seq(1,res@Number),ncol(partBinaryGE)),
+ nrow = res@Number,byrow = FALSE)
> (res@NumberxCol*1)*colMatrix
     [,1] [,2] [,3] [,4] [,5] [,6] [,7] [,8] [,9] [,10]
[1,]    0    0    1    1    1    0    0    0    0     0
[2,]    0    0    0    0    0    2    2    2    0     0
[3,]    0    0    0    0    0    3    3    0    0     3
[4,]    0    0    0    0    0    4    4    4    4     0
[5,]    0    0    0    5    5    0    0    0    0     5
[6,]    0    0    0    0    6    6    6    0    0     0
[7,]    0    0    0    0    7    7    7    7    0     0
[8,]    0    0    0    8    8    8    8    0    0     0
```

If we combine the bicluster assignments of the genes and those for the time points, then we will see some clusters overlap with each other. For instance, gene 6 and 7 show the same direction of expression from time points 6 to 9, and they are included in one bicluster (bicluster 4). However, gene 7 and time points 6 and 7 are also assigned to another bicluster formed by genes $2, 3, 4, 7, 8, 9$ and time points $5, 6, 7$ (bicluster 6). This type of ambiguity is helpful in the study of biological processes but sometimes can make it difficult when the goal is to detect biomarkers.

The plaid model biclustering The plaid model [89] identifies biclusters without dichotomizing the data and is more flexible in terms of inclusion of a row variable or a column variable into a cluster. A cluster can be a collection of rows and columns such that a certain number of rows and columns in the cluster share the same features. In addition, the cluster also includes a subset of those columns that may own such feature across all to-be-clustered rows, as well as a subset of the rows that have such feature across all columns. Under this context, Y_{ij} in a data matrix Y

with n rows and p columns is modeled as [140],

$$Y_{ij} = \Theta_{ij0} + \sum_{k=1}^{K} \Theta_{ijk} \rho_{ik} \kappa_{jk} + \epsilon_{ij}$$

$$= (\mu_0 + \alpha_{i0} + \beta_{j0} + \sum_{k=1}^{K} (\mu_k + \alpha_{ik} + \beta_{jk}) \rho_{ik} \kappa_{jk} + \epsilon_{ij},$$

where Θ_{ij0} denotes background features shared by all elements in the data, K represents the number of clusters (layers as noted in [89]), and μ_k denotes the feature shared by all elements in cluster k. Parameters α_{ik} and β_{jk} describe the effect of row i and column j, respectively. In this case, each row and each column in a data matrix have their unique features in addition to a common feature shared by all elements in cluster k. For instance, if entry Y_{12} is in cluster 1, then its unique feature is represented by $\mu_1 + \alpha_{11} + \beta_{21}$, where μ_1 is the same for all elements in cluster 1, α_{11} is shared by all elements in the first row of cluster 1, and β_{21} is for all elements in the second column in cluster 1. In the plaid model, α_{ik} and β_{jk} do not have to appear in all clusters. This flexibility is attractive due to its potential to fit different types of data and different needs of clustering [89]. As in the Bimax biclustering approach, the plaid model also allows overlapped biclusters.

To demonstrate this clustering method, we again use the normalized expressions data of the 9 genes at 10 time points. The goal is to group these genes such that their expression levels agree with each other across a set of consecutive time points, not just agreement in the direction of expression levels as in Bimax biclustering.

```
> matrixY = data.y[,1,]
> partMatrixY = matrixY[1:9,1:10]
> partMatrixY
        [,1]  [,2]  [,3]  [,4]  [,5]  [,6]  [,7]  [,8]  [,9] [,10]
[1,] -2.42 -2.15  0.66  1.98  1.55  0.78  0.14 -0.94 -1.03 -1.50
[2,] -0.35 -0.53 -0.13  0.01  0.37 -0.11  0.28 -0.20 -0.13 -0.14
[3,] -0.79 -0.25  1.18  0.84  0.19 -0.11 -0.42 -0.56 -0.36 -0.12
[4,] -0.28 -0.28  0.10  0.34  0.42  0.34  0.13 -0.33 -0.24 -0.06
[5,] -0.40 -0.64  0.32  0.39  0.27  0.62  0.68 -0.11 -0.13 -0.42
[6,] -1.15 -0.86  1.21  1.62  1.12  0.16 -0.44 -0.93 -1.23 -0.62
[7,] -0.52 -0.28 -0.65 -0.13 -0.15  0.34  0.33  0.26 -0.05 -0.03
[8,] -1.54 -0.14 -1.05 -1.16 -0.69  0.08  0.61  0.76  0.75  0.44
[9,] -0.45 -0.15  0.53  0.81  0.48  0.30  0.25 -0.23 -0.67 -0.51
```

To bicluster `partMatrixY`, the same function `biclust` is applied with arguments specified for the plaid model,

```
> res = biclust(x = partMatrixY, method = BCPlaid(),shuffle = 4,
+ verbose = FALSE)
> res@Number
[1] 2
```

In total, two clusters are identified. The default setting for the method BCPlaid() is to assume there is a background in the data, background=TRUE. The two clusters identified by the plaid model are the clusters identified after taking into account the background. To see the row clusters, we use the same approach as in Bimax,

```
> rowMatrix = matrix(rep(seq(1,res@Number), nrow(partMatrixY)),
+ ncol = res@Number,byrow = TRUE)
> (res@RowxNumber*1)*rowMatrix
      [,1] [,2]
 [1,]   1    0
 [2,]   0    0
 [3,]   0    0
 [4,]   0    0
 [5,]   0    0
 [6,]   1    0
 [7,]   0    2
 [8,]   0    2
 [9,]   0    0
```

The biclusters identified by the plaid model indicated that five genes do not belong to any clusters and their expressions represent the background, two genes (genes 1 and 6) are in bicluster 1, and another two genes (genes 7 and 8) are in bicluster 2. The cluster assignment of genes combined with the cluster assignment of time (10 time points) given below reveals the elements in each bicluster along with elements in the background; expressions of genes 1 and 6 at time points 4 and 5 are similar and different from the expressions in clusters 2 and expressions at the background, while expressions of genes 7 and 8 at time points 3 to 5 are similar and unique compared to expression of other genes.

```
> colMatrix<-matrix(rep(seq(1,res@Number), ncol(partMatrixY)),
+ nrow = res@Number, byrow = FALSE)
> (res@NumberxCol*1)*colMatrix
      [,1] [,2] [,3] [,4] [,5] [,6] [,7] [,8] [,9] [,10]
[1,]   0    0    0    1    1    0    0    0    0     0
[2,]   0    0    2    2    2    0    0    0    0     0
```

Compared to Bimax biclustering approach, the plaid model takes into account random variations in the data when biclustering. Thus this approach potentially gains a strength of inferring parsimonious biclusters that are practically more informative and interpretable.

The biclustering concept introduced in this section focuses on coherence of rows and columns in a data set. Essentially the technique is not model-based. Although the plaid model did assume a statistical model, it is all based on means and no associations are modeled. Thus these types of methods are restricted to profiles (e.g., means) in the variables and external variables do not play a role in the evaluation of similarity between different variables. Furthermore, some biclustering methods perform cluster analyses on the rows and columns separately, and do not simultaneously consider the interrelationship between the rows and columns. In addition, biclustering methods in general assume independence between clustering variables, which can potentially cause misclustering.

5.4 JOINT CLUSTER ANALYSIS

Different from bi-clustering, the concept of joint clustering, introduced in [162], is a probabilistic clustering approach and takes into account the correlations between variables (e.g., expression of genes in a biological pathway) and the interrelationships between variables and subjects. The process of clustering still focuses on coherence of rows and columns of a data set but with respect to an association with external variables, e.g., time.

The clusters are formed by consistent associations between a variable (or a "dependent variable") and covariates of interests among a subset of subjects for a set of variables (Figure 5.19). Each joint cluster is composed of a certain numbers of variables and a subset of subjects. Joint probabilistic clustering is relatively limited in the literature. In this section, we focus on the method proposed by Zhang et al. [162], which has an extension in a recent work by Han et al. [63]. Briefly, a semi-parametric model via penalized splines [36] is to evaluate possibly nonlinear associations between variables and covariates. To assign a variable to a specific cluster, a vector of indicators denoting cluster assignment of the variable is introduced. This vector is composed of zeros except for one location, and the length of the vector is determined by the number of variables to be clustered. To cluster subjects, a Dirichlet process mixture model is applied. The proposed joint clustering method has the ability to produce homogeneous clusters composed of a certain number of subjects sharing common features in the relationship between some variables and covariates [162]. In the following, we briefly present the ideas of the method.

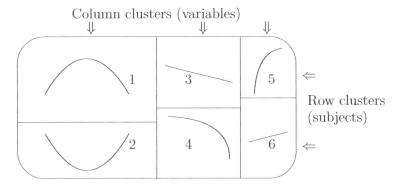

Column clusters (variables)

Row clusters (subjects)

Figure 5.19   Illustration of joint clusters. In total 6 clusters and the numbers are joint cluster indices.

Clustering the Variables   Clustering variables utilizes agreement in relationships between variables and covariates of interest. Assume in total n subjects and K variables are to be clustered. For subject $i, i = 1, \cdots, n$, let $\boldsymbol{y}_i = (y_{i1}, \ldots, y_{iK})$ denote the measures of K variables. Let M denote the number of clusters formed by the variables ($M \leq K$), and D be an $M \times K$ 0-1 matrix such that the kth column contains $M - 1$ zeros and one element with the number 1 indicating which cluster the kth variable ($k = 1, 2, \ldots, K$) belongs to. The value of M can be determined via grid search or pre-specified based on experience. For a given M, elements in the mth row of D, $m = 1, \cdots, M$, inform which variables are in cluster m. We formulate the variable clustering procedure into the following:

$$\boldsymbol{y}_{i,m} | D_{m.} = \boldsymbol{Q}(\boldsymbol{x}_i, \boldsymbol{\beta}_{i,m}) + \boldsymbol{\varepsilon}_{i,m}^T, \tag{5.2}$$

where $\boldsymbol{y}_{i,m} = (y_{i,(1)}, \ldots, y_{i,(k_m)})'$ is a vector of variables in variable cluster m, $\boldsymbol{x}_i$ is a vector of covariates potentially associated with $\boldsymbol{y}_{i,m}$, $\boldsymbol{\beta}_{i,m}$ describes the association of $\boldsymbol{y}_{i,m}$ with $\boldsymbol{x}_i$ in cluster variable m for subject i, and $\boldsymbol{\varepsilon}_{i,m}^T$ follows a multivariate normal distribution with mean $\boldsymbol{0}$ and covariance matrix Σ_m. The covariance Σ_m informs the strength of correlations between the variables in cluster m. Function $\boldsymbol{Q}(\cdot)$ is a vector function which describes the relationship between variables and covariates of interest $\boldsymbol{x}_i$. In [162], penalized splines (P-Splines) are applied [36] due to its use of low rank bases.

Clustering the Subjects   The subjects within each variable cluster are further grouped such that each group reflects a different relationship between variables and covariates of interest. Zhang et al. [162] proposes to use the Dirichlet process to cluster subjects [3]. The Dirichlet process has the ability to detect clusters without the need of defining a particular parameter for the number of clusters as done under the multinomial setting. Specifically, it is assumed that the prior distribution of $\beta_{i,m}$ is generated from a Dirichlet process,

$$
\begin{aligned}
\beta_{i,m}|G &\sim G, \\
G &\sim \mathrm{DP}(G_0, \lambda), \\
G_0|\sigma_0^2 &\sim N(0, \sigma_0^2 I), \\
\sigma_0^2|(a,c) &\sim \mathrm{InvGamma}(a,c),
\end{aligned}
$$

where $\beta_{i,m}|G$ are independent given G, and $\mathrm{DP}(G_0, \lambda)$ represents the Dirichlet process with a measure having concentration λ and proportional to the base distribution $G_0 \sim N(0, \Sigma_0)$ with $\Sigma_0 = \sigma_0^2 I$. The prior of $\beta_{i,m}$ conditional on $\beta_{-i,m}$, the coefficients with $\beta_{i,m}$ excluded, is a mixture distribution

$$
\beta_{i,m}\big|\beta_{-i,m} \sim \frac{1}{n-1+\lambda}\sum_{j\neq i}\delta_{\beta_{i,m}}(\beta_{j,m}) + \frac{\lambda}{n-1+\lambda}G_0,
$$

where $j = 1, \cdots, n, j \neq i$, $\delta_{\beta_{i,m}}(\beta_{j,m})$ is a point mass concentrated at a single point where $\beta_{i,m} = \beta_{j,m}$ (i.e., $\delta_{\beta_{i,m}}(\beta_{j,m}) = 1$ if $\beta_{i,m} = \beta_{j,m}, j \neq i$), and λ is the concentration parameter selected by maximizing joint posterior likelihood. Hyper-prior parameters a and c are assumed to be known and selected to achieve vague priors.

The above two processes are unified to jointly cluster variables and subjects:

$$
\begin{aligned}
y_{i,m}|(\beta_{i,m}, D, \Sigma_m) &\sim N(X_i'\beta_{i,m}, \Sigma_m), \\
\beta_{i,m}|G &\sim G, \quad G \sim \mathrm{DP}(G_0, \lambda), \\
G_0|\sigma_0^2 &\sim N(0, \sigma_0^2 I), \\
\sigma_0^2|(a,c) &\sim \mathrm{InvGamma}(a,c), \\
D_{.k}|\pi &\sim \mathrm{Multinomial}(1, \pi), \\
\pi|\zeta &\sim \mathrm{Dirichlet}(\zeta \mathbf{1}_M), \\
\zeta &\sim p(\zeta) = \frac{1}{2} \text{ if } 0 < \zeta \leq 1, \text{ and } \frac{1}{2}\zeta^{-2} \text{ if } \zeta > 1, \\
\Sigma_m|(S, \nu) &\sim \mathrm{InvWishart}(S, \nu),
\end{aligned}
$$

where $S = \frac{1}{2}\boldsymbol{I}$ with I standing for identity matrix and taking ν such that the prior mean of Σ_m is moderate on the diagonal.

Based on the above prior distributions, Zhang et al. [162] utilized Gibbs sampler to draw posterior samples and infer joint clusters clusters and patterns in each cluster. This approach has not been built into an R package, but the programs with examples are available at `https://www.memphis.edu/sph/contact/faculty_profiles/zhang.php`.

## CHAPTER 6

Methods to select genetic and epigenetic factors based on linear associations

In genetic and epigenetic studies, among a large number of candidate genes possibly associated with a health condition under study, it is desirable to identify important genes critical to disease risk. This type of effort is in the scope of variable selection problems. In this area, variable selection in linear or generalized linear regression models is often conducted. In addition to the methods built on Akaike information criterion (AIC) and Bayesian information criterion (BIC), more advanced variable selection methods have been proposed. In the following two subsections, we discuss methods under the frequentist framework and those built upon Bayesian inference. In all the variable selection methods, we consider the following classical linear regression model that has p candidate variables,

$$Y_i = \boldsymbol{X}_i^T \boldsymbol{\beta} + \epsilon_i, \ i = 1, \cdots, N,$$

where $\boldsymbol{\beta} = (\beta_1, \cdots, \beta_p)^T$ and $\epsilon_i \sim N(0, \sigma^2)$.

In the following sections, we introduce frequentist and Bayesian approaches, and demonstrate the methods using simulated data. In the last section, we apply the methods to DNA methylation data.

6.1 FREQUENTIST APPROACHES

We focus on three methods: elastic net, adaptive LASSO, and SCAD. Elastic net proposed by Zou and Hastie [166] is a regularization technique applied in variable selection and is a hybrid of LASSO [138] and ridge regression [67]. This technique is capable to appropriately select variables when predictors are correlated. The adaptive LASSO in [165] is an extension of LASSO and aimed to reduce the estimation bias in the LASSO developed by Tibshirani [138] to achieve the oracle properties, meaning that the method will correctly select the model as if the correct submodel were known. Another variable selection method also enjoying the oracle properties is the nonconcave penalized likelihood method developed by Fan and Li [37], where a smoothly clipped absolute deviation (SCAD) penalty is introduced. To estimate regression coefficients on important variables, these methods aim to solve the following penalized objective function,

$$\min_{\beta_0, \boldsymbol{\beta}} \frac{1}{N} \sum_{i=1}^{N} w_i l(y_i, \beta_0 + \boldsymbol{\beta}^T \boldsymbol{x}_i) + p(\boldsymbol{\beta}, \lambda, \cdot),$$

where w_i denotes weights and reflects the contribution of each observation, and $l(y_i, \beta_0 + \boldsymbol{\beta}^T \boldsymbol{x}_i)$ is the negative log-likelihood of each observation. When the distribution of $\boldsymbol{y}$ is Gaussian, $l(\cdot)$ is equivalent to a squared error of each observation. In the last term, $p(\boldsymbol{\beta}, \lambda, \cdot)$ is a penalty and is a function of $\boldsymbol{\beta}$ and other to-be-defined parameters represented by ".". In all these techniques, a regularization parameter λ needs to be specified, usually determined through a grid search via cross-validation.

6.1.1 Elastic net

The penalty in elastic net is defined as

$$
\begin{aligned}
p(\boldsymbol{\beta}, \lambda, \alpha) &= \lambda \Big\{ \big[(1 - \alpha) ||\boldsymbol{\beta}||_2^2 / 2 + \alpha ||\boldsymbol{\beta}||_1 \big] \Big\}, \\
&= \lambda \Big\{ \sum_{j=1}^{p} \big[(1 - \alpha) \beta_j^2 / 2 + \alpha |\boldsymbol{\beta}|_j \big] \Big\},
\end{aligned}
$$

where α is a mixing parameter. When $\alpha = 1$, $p(\boldsymbol{\beta}, 1)$ becomes the L_1 penalty in LASSO and when $\alpha = 0$, it is the L_2 penalty in ridge regressions. Since the penalty in LASSO penalizes the sum of absolute values of coefficients, when λ is large, many coefficients are shrunk to zeros, which never happens when the L_2 penalty as in ridge regressions is applied. Elastic net penalty, on the other hand, is a hybrid of these two types of techniques and it has been suggested that this approach can outperform LASSO with correlated candidate variables [166].

We have two parameters to tune in elastic net, α and λ, both of which can be tuned via re-sampling techniques. The R function `glmnet` in the package `glmnet` tunes λ for a fixed α, but it does not tune α. The R package `caret` is able to tune both α and λ through various re-sampling approaches such as cross-validation. It tests a range of possible α and λ values, and selects the best values for these two parameters which optimizes a statistic, e.g., prediction error (root mean square error) or deviance. To demonstrate the elastic net technique in variable selection, we first simulate a data set that has 10 candidate variables.

```
> library(MASS)
> sampsize = 200
> numVar = 10
> set.seed(12345)
> meanX = c(rep(2,3),rep(10,3),5,0.1,2, 0.8)
> varX = diag(c(rep(10,3),rep(5,3),5,2,1,2),length(meanX),
+ length(meanX))
> x = mvrnorm(sampsize,meanX,varX)
> # the second and third variables are important variables
> beta = c(0, 0.5, 1,rep(0,7))
> sig = diag(1,sampsize,sampsize)
> y = mvrnorm(1,x%*%beta,sig)
> data = cbind(y,x)
> colnames(data)<-c("y",paste('X',1:10,sep=""))
```

Of the 10 candidate variables, two variables are associated with the response variable with regression coefficients 0.5 and 1. Next, we use the `caret` package to tune the two parameters, α and λ, in elastic net and select important variables (i.e., variables with non-zero regression coefficients).

```
> library(caret)
> library(glmnet)
> set.seed(5000)
> # Define training control
> train_Control = trainControl(method = "cv",number = 10,
+                                  search = "random")
```

The function `trainControl` is to set up the control parameters to be used in the training process to tune α and λ. The argument `method` is to specify a re-sampling approach. A number of approaches are available and commonly used approaches are cv for cross-validation and repeadedcv for Monte-Carlo cross-validation or repeated random sub-sampling validation. Two options are available for search, grid and random, to specify the parameter space for the purpose of identifying the best values of α and λ. Compared to grid, the option random covers the parameter space to a lesser extent. In this example, the re-sampling method is a 10-fold cross-validation indicated by method = "cv" and number = 10. Next, based on the selected methods and parameters, we start the training process using the function train.

```
> TuneParms = train(y~., data = data, method = "glmnet",
+           trControl = train_Control, tuneLength = 60)
> # Best tuning parameters
> TuneParms$bestTune
       alpha     lambda
56 0.8972444 0.1311761
```

The method argument specifies that elastic net is used in this tuning process. In our example, random search instead of grid search is used for candidate values of α and λ, as defined in trainControl, the setting tuneLength=60 in this case defines the maximum number of tuning parameter combinations (which is 60) that will be generated by the random search. On the other hand, if grid search is used, then 60 is the number of levels for each tuning parameter to be generated. In our example, the best tuning parameters were reaching at the 56-th combination, which are 0.90 and 0.13.

Data of the predictors could also be pre-processed by use of the preProcess argument in train. For instance, to standardize the predictors to eliminate unit-dependence, we set preProcess=c("center", "scale").

The coefficients of the selected variables based on the tuned α and λ are extracted by coef. In total, two variables are selected, consistent with the underlying true model.

```
> coef(TuneParms$finalModel, TuneParms$bestTune$lambda)
11 x 1 sparse Matrix of class "dgCMatrix"
                     1
(Intercept) 0.05154811
X1             .
X2          0.46135487
X3          0.99471555
X4             .
X5             .
```

```
X6          .
X7          .
X8          .
X9          .
X10         .
```

Note that this process does not provide a statistical significance of each selected variable. To infer the statistical significance, a commonly chosen approach is to select the variables using a subset of data (assuming the sample size is larger than the number of candidate variables), and then use the remaining data to fit a linear regression model, e.g., using lm, to infer the statistical significance of each variable.

6.1.2 Adaptive LASSO

The approach of adaptive LASSO was proposed by Zou [165] and it was evolved from LASSO. Compared to LASSO, adaptive LASSO has the oracle properties under a proper choice of λ. The penalty in adaptive LASSO is defined as

$$p(\boldsymbol{\beta}, \lambda, \gamma) = \lambda \sum_{j=1}^{p} \hat{\omega}_j |\beta_j|,$$

$$\hat{\omega}_j = 1/|\hat{\beta}_j|^{\gamma},$$

where $\hat{\beta}_j$ is the estimate of regression coefficients based on ordinary least squares or from the best ridge regression in the situation of collinearity. We use the same simulated data to demonstrate this approach, starting from inferring $\hat{\omega}_j$ based on the best ridge regression via a 10-fold cross-validation.

```
> # Use ridge regression to construct the weights
> best_ridge<- cv.glmnet(x = x, y = y, type.measure = "mse",
+                        nfold = 10, alpha = 0)
> lambda_tuned = best_ridge$lambda.min
> lambda_tuned
[1] 0.3551413
> best_ridge_coef = as.numeric(coef(best_ridge,
                   s = lambda_tuned))[-1]
> # the weights, omega, calculated based on gamma=1
> omega = 1 / abs(best_ridge_coef)
> omega
 [1] 9978.806663     2.206854     1.056506    24.349330    23.465679
 [6]   21.591722   108.207493    71.379271    18.952099    15.469180
```

The parameter λ is tuned using 10-fold cross-validation in ridge regression using the R function `cv.glmnet`. The statistics mean squared error (`mse`) assesses the loss and is used to choose the best λ, and the value of λ corresponding to the smallest `mse` is 0.36. Next, ω is estimated as $1/|\hat{\beta}_j|$ setting $\gamma = 1$ by first extracting the estimated coefficients (excluding the intercept) using `coef` and then calculating ω as `omega = 1 / abs(best_ridge_coef)`. Based on the findings in [165], it seems taking $\gamma = 1$ overall works reasonably well.

Finally, we apply the weights and utilize `glmnet` to select informative variables and infer the coefficients of those variables.

```
> # use glmnet with the calculated adaptive LASSO weights, omega,
> # to perform variable selection.
> adaLASSO = cv.glmnet(x = x, y = y,type.measure = "mse",nfold = 10,
+                      alpha = 1,penalty.factor = omega)
> plot(adaLASSO,xlab = expression(paste("log(",lambda,")")))
> adaLASSO$lambda.min
[1] 10.78594
> coef(adaLASSO, adaLASSO$lambda.min)
11 x 1 sparse Matrix of class "dgCMatrix"
                     1
(Intercept) -0.09148639
V1           .
V2           0.49119666
V3           1.02828333
V4           .
V5           .
V6           .
V7           .
V8           .
V9           .
V10          .
```

In the above, a 10-fold cross-validation is used under the adaptive LASSO penalty specified by `alpha=1` and `penalty.factor=omega`. The tuned λ by minimizing `mse` using cross-validation is 10.79 (Figure 6.1). The selected variables using adaptive LASSO are consistent with the truth and the estimated coefficients are closer to the true values, compared to the results from elastic net.

6.1.3 Smoothly clipped absolute deviation (SCAD)

The smoothly clipped absolute deviation (SCAD) penalty is proposed by Fan and Li [37] and its derivative is given by

$$p'(\beta, \lambda, a) = \lambda \left\{ I(\beta_j \le \lambda) + \frac{(a\lambda - \beta_j)_+}{(a-1)\lambda} I(\beta_j > \lambda) \right\},$$

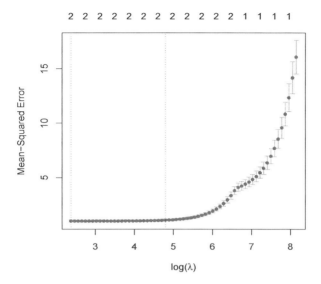

Figure 6.1   Mean square errors vs. $\log\lambda$ in cross-validation to tune λ in adaptive LASSO.

which leads to

$$
\hat{\beta}_j^{SCAD} = \begin{cases} (|\hat{\beta}_j| - \lambda)_+ \text{sign}(\hat{\beta}_j), & \text{if } |\hat{\beta}_j| \leq 2\lambda, \\ \left\{(a-1)|\hat{\beta}_j| - \text{sign}(\hat{\beta}_j)a\lambda\right\}/(a-2), & \text{if } 2\lambda < |\hat{\beta}_j| \leq a\lambda, \\ \hat{\beta}_j, & \text{if } |\hat{\beta}_j| > a\lambda, \end{cases}
$$

$j = 1, \cdots, p$, where Fan and Li [37] suggested to set $a = 3.7$. The SCAD penalty is continuously differentiable from $-\infty$ to $+\infty$ except at 0, resulting in small coefficients being shrunk to zero with large coefficients un-penalized. The SCAD penalty, as the adaptive LASSO, also enjoys the oracle property [37]. To apply the SCAD penalty in variable selection, the R package ncvreg can be applied. Again, we use the simulated data discussed earlier to demonstrate this approach using ncvreg.

```
> library(ncvreg)
> SCAD = cv.ncvreg(x, y, penalty = "SCAD", family = "gaussian",
+ nfolds = 10, alpha = 1, seed = 12345)
> plot(SCAD)
> SCAD$lambda.min
[1] 0.3551413
> SCAD$fit$beta[,SCAD$min]
```

```
(Intercept)             V1            V2            V3            V4
 -0.1134016     0.0000000     0.4984394     1.0312255     0.0000000
         V5            V6            V7            V8            V9
  0.0000000     0.0000000     0.0000000     0.0000000     0.0000000
        V10
  0.0000000
```

As in the other methods, a 10-fold cross-validation is applied to tune λ. Argument `alpha` is set at 1 to indicate the penalty is L_1 penalty. The best λ is selected as 0.36 by minimizing deviance (Figure 6.2), based on which variables X2 and X3 are selected with estimated parameters close to the truth.

6.2 BAYESIAN APPROACHES

Bayesian methods estimate the posterior probabilities for all models under consideration, rather than concluding a single model as done by frequentist approaches. A key component in variable selection under the Bayesian framework is the specification of prior distributions for the regression coefficients. Most work [52, 50, 51, 49, 48, 130, 39, 90, 74, 75, 159] in this area can be categorized into two groups, the g-prior initiated in [156] and the spike and slab models in [110]. In this section, we

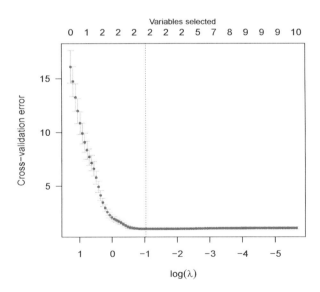

Figure 6.2 Cross-validation error (deviance) vs. $\log\lambda$ in to tune λ in SCAD.

discuss several examples in the scope of g-prior originally proposed by Zellner [156], its extension named as two-component G-prior [159], and the spike and slab model [52].

6.2.1 Zellner's g-prior

Zellner's g-prior is defined as

$$\boldsymbol{\beta}|g, \sigma^2 \quad \sim \quad N(0, g(\boldsymbol{X}^T\boldsymbol{X})^{-1}\sigma^2), \tag{6.1}$$

where g is unknown and needs to be specified [156]. Various approaches have been proposed to select g, for instance, AIC, BIC, or risk inflation criterion (RIC) by Foster and George [43] to calibrate g, empirical Bayes to select g [49], or inferring g by introducing hyper prior distributions to g [97]. Fernández et al. [39] compared a variety of choices of g and recommended choosing g as $\max(\sqrt{n}, p^2)$. In all these methods, the functionality of g is fulfilled through its influence on the variance components in the prior distribution of non-zero coefficients $\boldsymbol{\beta}$.

To apply the Zellner's g-prior in variable selections, an R function GibbsBvs in the package BayesVarSel can be applied for this purpose and this package is suitable for large number of candidate variables, e.g., > 20. Different approaches in GibbsBvs are available to select g include setting g as the sample size [156] (gZellner), the multivariate Cauchy prior for g [157, 155] (ZellnerSiow), the utilization of hyperprior of g [97] (Liangetal), the benchmark prior suggested by Fernández et al. [39] (FLS), and a robust approach proposed by Bayarri et al. [6] (Robust). As for the priors to the models (prior.models), three choices are available, prior.models="Constant", "ScottBerger", and "User". With prior.models="Constant", a priori, each possible model has the same probability to be the true underlying model. With setting "User", the probability of each model is defined by priorprobs such that the first probability corresponding to the simplest model defined by fixed.cov in the function GibbsBvs, and each subsequent model is with one more covariate added until the most complex model. If in total p candidate variables, then the size of vector priorprobs is $p + 1$. Finally, taking prior.models="ScottBerger" is the same as taking prior.models="User" and priorprobs = 1/choose(p,0:p), where choose(p,0:p) represents the combination calculations, e.g., choose(5,2)=10. The advantage of using ScottBerger is its ability to adjust multiple testing in variable selections. The sampling approach in GibbsBvS is the Gibbs sampler, which chooses top models based on frequencies of visits [50] instead of inferring the posterior probabilities of each model.

In our illustration, we focus on four methods used to define or model g, gZellner, Liangetal, FLS and Robust and apply the package to our simulated data to examine how each of these methods performs.

```
> # Zellner's g prior
> library(BayesVarSel)
> dataFrame<-as.data.frame(data)
>
> # gZellner
> SimuGibbsBvs = GibbsBvs(formula = y ~ .,data = dataFrame,
+          prior.betas = "gZellner", prior.models = "Constant",
+          n.iter = 10000, init.model = "Full", n.burnin = 5000,
+          time.test = FALSE)
Info. . . .
Most complex model has 11 covariates
From those 1 is fixed and we should select from the remaining 10
X1, X2, X3, X4, X5, X6, X7, X8, X9, X10
The problem has a total of 1024 competing models
Of these, 15000 are sampled with replacement
Then, 10000 are kept and used to construct the summaries
Working on the problem...please wait.
Warning message:
In GibbsBvs(formula = y ~ ., data = dataFrame, prior.betas =
"gZellner",  :
  The number of variables is small enough to visit every model.
  Consider Bvs (or pBvs for its parallel version).
```

In the above, the value of g is the sample size (200 in this simulation example), as suggested by Zellner [156]. Some information is provided after running the line. The null model only has the intercept. Since we have 10 covariates, in total, we select from $2^{10} = 1,024$ models. The number of burn-in iterations is n.burnin=5,000. After 5,000 iterations, additional 10,000 iterations are run and used to draw posterior inferences. Since the number of variables is relatively small, in the warning message, function Bvs is suggested, in which case a posterior probability of each model is inferred. Function Bvs essentially has the same arguments as GibbsBvs, except that Bvs allows parallel computing.

```
> SimuGibbsBvs
```

```
Call:
```

```
GibbsBvs(formula = y ~ ., data = dataFrame,
    prior.betas = "gZellner", prior.models = "Constant",
    n.iter = 10000, init.model = "Full",
    n.burnin = 5000, time.test = FALSE)
```

```
The 10 most probable models among the visited ones are:
   X1 X2 X3 X4 X5 X6 X7 X8 X9 X10
1      *  *
2      *  *  *
3      *  *                    *
4      *  *        *
5      *  *              *
6      *  *     *
7      *  *                 *
8      *  *           *
9   *  *  *
10     *  *  *                 *
> SimuGibbsBvs$inclprob
    X1      X2      X3      X4      X5      X6      X7
0.0668  1.0000  1.0000  0.1444  0.0805  0.1204  0.0659
    X8      X9     X10
0.0944  0.0765  0.1280
```

The results include the top 10 models with the highest visiting frequencies and the model with variables X2 and X3 are the best. The inferred posterior probabilties of inclusion for each variable also indicate that X3 and X4 are the most important ones.

For this simulated data, the top models selected based on FLS and Robust are the same as the models chosen by taking prior.betas = "gZellner". However, when setting prior.betas = "Liangetal", an error message was received,

```
Error in GibbsBvs(formula = y ~ ., data = dataFrame,
prior.betas = "Liangetal",  :
  A Bayes factor is infinite.
```

This error will occur if the sample size is relatively large or a model is much better than the null model. Currently, this problem has not been fixed, and it is suggested to use gZellner when such a problem happens.

6.2.2 Extension of Zellner's g-prior to multi-components G-prior

The selection of g in Zellner's g-prior is critical; overly big or small values of g both will lead to exclusion of important variables [100]. This

Lindley's paradox of Zellner's g-prior in variable selection confesses the importance of proper choice of g [100, 5]. However, a single g is a global shrinkage factor applied to all predictors and will equally shrink each coordinate. This choice is reasonable if we do not know anything about the predictors or if they are equivalent in some sense as noted in [132]. Otherwise, multi-components G reflecting different levels of importance of different variables are necessary. We introduce a published method identified so far with such extensions, the two-components G-prior [159], and would like to note that a couple of studies posted on arXiv are identified which further extended to more than two-components G-priors, e.g., [132, 104].

In the two-components G-prior, data are essentially categorized to two groups, informative and non-informative a priori. Accordingly, for each variable $x_j, j = 1, \cdots, p$, the prior distribution of its corresponding regression coefficient is taken as the Zellner's g-prior with $g = g_j$. This leads to the definition of the G, $G = \text{diag}(g_1, \cdots, g_p)$, a diagonal matrix composed of g_j's. A prior point mass function on g_j is defined as $p(g_j) = q_j I(\{g_j = g_l\}) + (1 - q_j) I(\{g_j = g_s\})$, where $I(\cdot)$ is an indicator variable and $I(\cdot) = 1$ if $(\cdot)$ is true, q_j is the probability of $g_j = g_l$, and $g_l, g_s > 0$ such that $g_l = b f_1(n), g_s = b f_2(n)$ with $f_1(n) = O(n), f_2(n) = O(n^\psi)$, $1/2 < \psi < 1$, implying $f_2(n) = o(f_1(n))$ as $n \to \infty$. Based on this definition, we have, as $n \to \infty$, $g_s \to \infty$, $g_l \to \infty$, but $g_s/g_l \to 0$.

The two functions $f_1(n)$ and $f_2(n)$ in the definition of g_l and g_s are assumed to be known, e.g., to satisfy the requirement, we can choose $f_1(n) = n$, and $f_2(n) = n^{0.55}$. The definitions of $f_1(n)$ and $f_2(n)$ follow the generic suggestion for the choice of g as stated in [39]. The parameter $b > 1$ is a tuning parameter and determines the distance between g_l and g_s, for a given $f_1(n)$ and $f_2(n)$. With the utilization of G, the prior distribution of β is the so-named two-component G-prior, as defined in [159],

$$
\begin{aligned}
\beta_\gamma | X_\gamma, \gamma, \sigma^2, G \quad &\sim \quad N\left\{0, \sigma^2 \left[\left(X_\gamma G_\gamma^{-1}\right)^T \left(X_\gamma G_\gamma^{-1}\right)\right]^{-1}\right\}, \\
&= \quad N\left\{0, \sigma^2 G_\gamma \left(X_\gamma^T X_\gamma\right)^{-1} G_\gamma\right\}
\end{aligned}
\tag{6.2}
$$

where γ is a vector of length p composed of 1s and 0s indicating the inclusion or exclusion of a variable, respectively. Matrix X_γ is an $n \times \Lambda_\gamma$ model matrix with columns chosen by γ, where $\Lambda_\gamma = \gamma^T \mathbf{1}$ is the size of a model determined by γ. Next, G_γ is an $\Lambda_\gamma \times \Lambda_\gamma$ diagonal matrix. Note that $\boldsymbol{\beta_\gamma}$ instead of $\boldsymbol{\beta}$ is used in (6.2) to indicate the dependence of $\boldsymbol{\beta}$ on γ. Only coefficients of selected variables are in $\boldsymbol{\beta_\gamma}$, and the coefficients of unselected variables are zeros with probability 1 [50].

Note that if $g_j = g_s$ or $g_j = g_l$ for all $j = 1, \cdots, p$, then the prior is essentially the Zellner's g-prior. The ultimate functionality of b in (6.2) is to optimize g_j to eliminate unimportant variables and thus a careful consideration in the selection b is in great need. Zhang et al. [159] proposed an effective approach to select b utilizing pseudo variables. As shown and demonstrated in [159], a nice property of this two-components $\boldsymbol{G}$-prior is its efficiency in identifying truly important variables, compared to the original g-prior.

6.2.3 The spike-and-slab prior

The spike-and-slab prior distribution for coefficients in regressions was first introduced in [110] such that the prior is a mixture of two components, a point mass function at zero (the spike component) and a flat prior distribution (the slab component). The prior distributions for regression coefficients later proposed in [52, 74, 75] are extension of the original spike-and-slab prior. In particular, in stead of using point mass function for the spike component, George and McCulloch [52] proposed a mixture of two normal distributions representing respectively the spike-and-slab parts, which is included later in a review article by the same authors [50],

$$\beta_j | \gamma_j \quad \sim \quad (1 - \gamma_j)N(0, \tau_j^2) + \gamma_j N(0, c_j^2 \tau_j^2), \qquad (6.3)$$

where γ_j is an indicator variable taking values 0 or 1. In [52], τ_j is set small such that when $\gamma_j = 0$, then β_j would be small enough to be treated as 0, thus excluding variable X_j. On the other hand, c_j is set large with $c_j > 1$ such that when $\gamma_j = 1$, then a non-zero estimate of β_j is suggested and thus variable X_j is included in the model. George and McCulloch [52] gives suggestions on the selection of τ_j and c_j, for instance, taking τ_j similar to the standard error of the least squares estimate of β_j and $c_j = 10$. Compared to the spike-and-slab prior in [110], this prior distribution enjoys great computational efficiency.

This spike-and-slab prior is programmed into an R function, `lm.spike`, available in the `BoomSpikeSlab` package. We again use the example data to demonstrate the application of this function in selecting important variables using the prior defined in (6.3).

```
> library(BoomSpikeSlab)
```

```
> niter = 1000
> model = lm.spike(y ~ x, niter = niter,ping = (niter/2-1))
=-=-=-=-= Iteration 0 Thu Aug 22 22:02:37 2019
 =-=-=-=-=
=-=-=-=-= Iteration 499 Thu Aug 22 22:02:37 2019
 =-=-=-=-=
=-=-=-=-= Iteration 998 Thu Aug 22 22:02:37 2019
 =-=-=-=-=
```

In the function `lm.spike`, the prior distributions for the parameters including selection of hyper-prior parameters are defined in `SpikeSlabPrior`. In this example, all hyper-prior parameters are taken as the default set in `SpikeSlabPrior`. We use `plot.ts` to draw trace plots, which help us to examine the convergence of posterior sampling of the regression coefficients β (Figure 6.3). We can do the same for the variance parameter, σ^2, in the normal distribution assumption for model random errors. From the trace plots, the convergence of the MCMC chains for β can be reasonably induced, and we can infer that the coefficients of variables X2 and X3 are non-zero.

```
> plot.ts(model$beta[,-1],main = expression(paste(
+ "Trace plots for ",beta)))
```

To examine the statistics of variable selection such as selection probabilities, we use the `summary` function as shown below. The data in the first two columns give the estimated posterior mean and standard deviation of each regression coefficients and the last column is for the selection probability of each variable. The probabilities of including X2 and X3 are both 1, while other variables are with rather low probability to be included in the model. The posterior mean for the coefficients of X2 and X3 are both close to the true values (1 and 0.5, respectively). Summary statistics, `mean.inc` and `sd.inc`, in the remaining two columns are the means and standard deviations of the variables when they are included in the model in the MCMC process.

```
> summary(model)
coefficients:
              mean          sd mean.inc  sd.inc inc.prob
x3        1.02e+00 0.018100   1.0200 0.01810    1.000
x2        4.87e-01 0.020300   0.4870 0.02030    1.000
x5       -8.50e-05 0.001400  -0.0170 0.01150    0.005
x4       -8.02e-05 0.001240  -0.0160 0.00797    0.005
x9       -9.88e-05 0.002320  -0.0329 0.03270    0.003
```

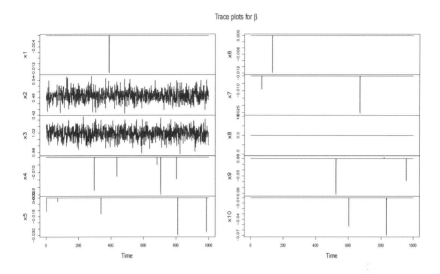

Figure 6.3 Trace plots of the MCMC posterior sampling of regression coefficients β.

```
x10            -1.24e-04 0.002790  -0.0619 0.01160   0.002
x7             -3.48e-05 0.000861  -0.0174 0.01170   0.002
(Intercept) -2.03e-04 0.004730  -0.1010 0.04340   0.002
x6             -1.19e-05 0.000378  -0.0119 0.00000   0.001
x1             -1.34e-05 0.000423  -0.0134 0.00000   0.001
x8              0.00e+00 0.000000   0.0000 0.00000   0.000


residual.sd =
   Min. 1st Qu.  Median    Mean 3rd Qu.    Max.
 0.8624  0.9502  0.9825  0.9867  1.0202  1.1490


r-square     =
   Min. 1st Qu.  Median    Mean 3rd Qu.    Max.
 0.9179  0.9353  0.9400  0.9393  0.9439  0.9538
```

In addition to utilizing spike-and-slab priors on regression coefficients, Ishwaran and Rao [74, 75] implemented the idea of spike-and-slab to the scale parameters in the prior distribution of regression coefficients. The prior distribution of β_j is normal with mean zero and variance $I_j \tau_j^2$,

$$\beta_j | \gamma_j \quad \sim \quad N(0, I_j \tau_j^2).$$

The prior distribution of I_j is a spike-and-slab prior,

$$
\begin{aligned}
I_j &\sim (1 - w_k)\delta_{v_0}(\cdot) + w_k\delta_1(\cdot), \\
w_j &\sim \text{uniform}[0, 1],
\end{aligned} \tag{6.4}
$$

where the value of v_0 is small near 0. The prior distribution of τ_j^2 is assumed to be an inverse gamma distribution with shape and scale parameter selected so that the distribution of $I_j\tau_j^2$ is a continuous bimodal with a spike at v_0 and a right-continuous tail (Figure 7 in [74]). Ishwaran and Rao also proposed to re-scale the response variable to control the amount of penalization and encourage selective shrinkage. In another work of the same research group [76], they extended the re-scaled spike-and-slab prior to fit the situation when sample size is smaller than the number of variables, which is common in genetic and epigenetic studies. This approach selects variables by generalizing the elastic net penalty proposed by Zou and Hastie [166] with regression parameters estimated based on the re-scaled spike-and-slab prior.

The algorithm has three steps in order, including filtering to exclude unimportant variables, Bayesian model averaging (BMA) to obtain posterior mean of the regression coefficients, and selecting variables based on generalized elastic net (*gnet*) with regression coefficients estimated from BMA. The re-scaled spike-and-slab prior is utilized in both the screening step and the BMA step. In the filtering step, a pre-specified number of variables of interest are kept based on posterior means of the regression coefficients estimated by use of the spike-and-slab prior. In the BMA step, the spike-and-slab prior is applied to those variables that pass the screening in the first step and posterior mean of the regression coefficients are estimated. These estimated regression coefficients are then utilized in the last step to select variables via the generalized elastic net. The algorithm is built into an R package `spikeslab` [77]. In the following, we apply the R function `spikeslab` to the same simulated data discussed in earlier sessions of this chapter.

```
> library(spikeslab)
> model = spikeslab(y ~ x)
```

The regression model set up is the same as before, $y \sim x$. Parameters in the prior distributions are chosen as the default values suggested by the package [77]. In addition, by default, the response y is centered and the predictors in x are standardized to mean 0 and variance of 1. They emphasize the importance of centralization and standardization and strongly suggest taking the default setting in the function, which is likely due to the implementation of the re-scaled spike-and-slab prior [74].

To find out other settings in the variable selection process, we use print,

```
> print(model)
--------------------------------------------------------------------
Variable selection method      : AIC
Big p small n                  : FALSE
Screen variables               : FALSE
Fast processing                : TRUE
Sample size                    : 200
No. predictors                 : 10
No. burn-in values             : 500
No. sampled values             : 500
Estimated mse                  : 0.9799
Model size                     : 2


---> Top variables:
     bma   gnet bma.scale gnet.scale
x3 3.456 3.389     1.028      1.008
x2 1.526 1.478     0.488      0.473
--------------------------------------------------------------------
> plot(model)
```

The output from print(model) shows that the final variable selection step using *gnet* is determined by AIC. As discussed in Ishwaran et al. [77], due to the inclusion of BMA, the *gnet* estimator is in general stable and a simple criterion such as AIC is sufficient in the path of optimization. Furthermore, since in the simulated data, the number of variables is smaller than the sample size, screening is not used (bigp.smalln = FALSE and screen = FALSE). As shown in the second part of the output, the two important variables, X2, X3, are correctly selected with the estimated regression coefficients close to the true values. Note that the column labeled as gnet includes coefficients with standardized predictors and gnet.scale are the coefficients re-scaled back to the original predictors. The same logic is applied to the columns bma and bma.scale. The estimated coefficients in the column gnet.scale are consistent with the true values. Finally, we visualize the optimization process by use of plot(model) (Figure 6.4), which clearly indicate the correct selection of variables X2 and X3.

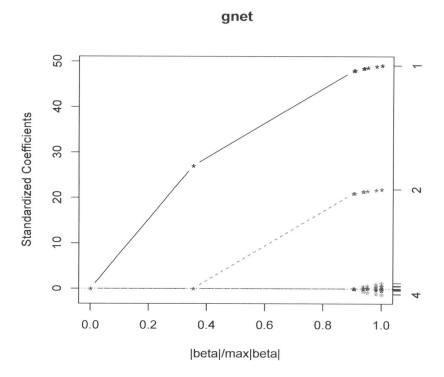

Figure 6.4 Plot of the optimization path of gnet in the process of variable selection.

6.3 EXAMPLES – SELECTING IMPORTANT EPIGENETIC FAC- TORS

We finish this chapter by demonstrating the methods discussed in this chapter using DNA methylation data provided by the R package `DMRcate`. We focus on the methods with their R packages available. This data set contains DNA methylation of 38 matched pairs with 38 tumor samples. We first implement the `ttScreening` package to identify a CpG site such that DNA methylation at this CpG site is most likely to be associated with tumor status and treat this CpG site as a marker site. Next through variable selections, we identify CpGs such that their DNA methylation is associated with DNA methylation at this marker site. In the following, we start from calling in and cleaning the data.

```
> library(DMRcate)
> data(dmrcatedata)
> DNAm = logit2(myBetas)
> dim(DNAm)
```

```
[1] 10042     76
> DNAmNoProbSNPs = rmSNPandCH(DNAm, dist=10, mafcut=0.05)
> dim(DNAmNoProbSNPs)
[1] 9126    76
> CpGs = rownames(DNAmNoProbSNPs)
> ID = factor(sub("-.*", "", colnames(DNAm)))
> Status = factor(sub(".*-", "", colnames(DNAm)))
```

After linking to the DMRcate library, we load the DNA methylation data by data(dmrcatedata). As introduced in Chapter 1, DNA methylation level is calculated as the proportions of intensity of methylated over the sum of methylated and unmethylated sites, named as β values. The logit with base 2 transformation is to convert DNA methylation to M values for the purpose of reducing heteroscedasticity as suggested by Du et al. [28]. This data set includes DNA methylation at $10,042$ CpG sites of 76 samples. The last three lines are to extract the names of the CpG sites, subject IDs, and tumor status of each individual. After data called in, next we screen CpG sites potentially associated with tumor status using the R package ttScreening.

```
> library(ttScreening)
> runs = ttScreening(y = DNAmNoProbSNPs, formula = ~Status,
+ data = Status, imp.var = 1,
+ sva.method = "two-step",
+ B.values = FALSE, iterations = 100, cv.cutoff = 50,
+ n.sv = NULL, train.alpha = 0.05, test.alpha = 0.05,
+ FDR.alpha = 0.05, Bon.alpha = 0.05, percent = (2/3),
+ linear = "ls", vfilter = NULL, B = 5,
+ numSVmethod = "be", rowname = NULL,maxit = 20)
Number of significant surrogate variables is:  22
[1] "9 surrogate variables used in the analysis"
> out = runs$TT.output
> dim(out)
[1] 667   24
```

In the above, we take the default values for all the arguments. In total, 667 CpGs are statistically significant at the 0.05 significance level in at least 50 pairs of training-testing data sets. We now identify the CpG site such that it has the highest credibility to serve as a biomarker of the disease. This is determined by frequency that a CpG site is statistically significant in both training and testing data sets.

```
> maxFreq = max(out[,2])
> maxFreq
```

```
[1] 100
> rowMaxFreq = as.numeric(as.character((out[which(out[,2]
+ ==maxFreq),1]))))
> y = t(DNAmNoProbSNPs[rowMaxFreq,])
> rownames(y) = CpGs[rowMaxFreq]
> rownames(y)
[1] "cg00493554"
> x = t(DNAmNoProbSNPs[-rowMaxFreq,])
> colnames(x) = CpGs[-rowMaxFreq]
```

As seen in the results from the codes above, cg00493554 is statistically significant across all the 100 randomly training and testing pairs. We will use this CpG site as the most promising biomarker for the disease. Our next step is to select CpG sites with DNA methylation associated with DNA methylation at this biomarker site. Since many of the approaches introduced in this chapter do not directly deal with the problem of large p (variables) small n (sample size), for the purpose of illustration, we include the first 20 variables in the illustration of all the variable selection methods except for the method in spikeslab.

```
> newx = x[,1:20]
> Data = cbind(t(y),newx)
```

In the following, we use the elastic net penalty to demonstrate the selection of CpG sites. The program is similar to what was presented earlier based on simulated data with minor changes. Thus, for other approaches introduced in this chapter, readers are expected to figure out with straight forward revisions to the codes. For the elastic net penalty, we use the grid search, instead of random search as used in earlier examples based on simulated data, to choose the values for α and λ. The name of the dependent variable in the function train in this real data example is cg00493554, the CpG site with the strongest potential to be a biomarker for the disease.

```
> # Elastic net

> set.seed(500)
> train_Control = trainControl(method = "cv",number = 10,
+                                 verboseIter = FALSE)
> # Tune the two parameters
> lambda.grid = seq(-2,2,length=50)
> alpha.grid = seq(0,1,length=10)
> srchGrd = expand.grid(.alpha = alpha.grid,.lambda = lambda.grid)
>
> TuneParms = train(cg00493554~., data = Data, method =
```

```
+            "glmnet",tuneGrid = srchGrd,
+            trControl = train_Control)
Warning message:
In nominalTrainWorkflow(x = x, y = y, wts = weights,
info = trainInfo,  :
  There were missing values in resampled performance measures.
>
> # Best tuning parameters
> TuneParms$bestTune
        alpha    lambda
178 0.3333333 0.2040816
```

On the grid, we evenly spaced 10 α values from 0 to 1 and 50 λ values from -2 to 2. Note that in elastic net penalty, unlike ridge regression and LASSO, the values of α is flexible and ranged between 0 and 1. The warning message from `train` is due to missing values at some grid values of α and λ. The best values identified for α and λ are 0.33 and 0.20, respectively.

We now extract the coefficients estimated based the chosen α and λ, using `coef`. These estimated coefficients are for the selected CpG sites such that DNA methylation at those CpGs are associated with DNA methylation at cg00493554.

```
> loc = which(coef(TuneParms$finalModel,
+     TuneParms$bestTune$lambda) != 0)
> names = row.names(coef(TuneParms$finalModel,
+     TuneParms$bestTune$lambda))
> names[loc[-1]]
[1] "cg00156569" "cg00157199" "cg00188822" "cg00226306"
[5] "cg00246386" "cg00254133" "cg00298324" "cg00309582"
> coef(TuneParms$finalModel, TuneParms$bestTune$lambda)[loc[-1]]
[1] -0.38905072 -0.21099692  0.05072564  0.46776298 -0.11194120
[6] -0.47126338  0.51944529  0.07505822
```

By use of elastic net penalty, we selected eight CpG sites associated with cg00493554 in DNA methylation. As a reference, we include a list of the selected CpGs and their regression coefficients for each of the approaches in Table 6.1. Note that cg00254133 and cg00298324 are selected by all the methods introduced so far.

Recall that the function `spikeslab` has the ability to select variables such that the number of candidate variables is larger than the sample size. In the epigenetic data, after excluding problematic CpG sites, DNA methylation at $9,126$ CpG sites was measured for each of the 76 subjects. Assuming at most 76 CpGs where DNA methylation is associated with the DNA methylation at the potential marker cg00493554, we apply `spikeslab` to select most important variables.

The argument `bigp.smalln` in `spikeslab` is set at `TRUE` indicating that we are dealing with a situation such that the number of variables is larger than the sample size. We set `bigp.smalln.factor = 1` meaning the maximum number of variables to be included in the model is $bigp.smalln.factor \times n = n$ with n being the sample size.

```
> model = spikeslab(t(y) ~ x , bigp.smalln = TRUE,
+          bigp.smalln.factor = 1,center=TRUE)
> print(model)
-----------------------------------------------------------------
Variable selection method    : AIC
Big p small n                : TRUE
Screen variables             : TRUE
Fast processing              : TRUE
Sample size                  : 76
No. predictors               : 9125
No. burn-in values           : 500
No. sampled values           : 500
Estimated mse                : 0.1044
Model size                   : 22


---> Top variables:
              bma    gnet bma.scale gnet.scale
xcg03770410  0.513  0.439     0.574      0.492
xcg17978103  0.509  0.484     0.401      0.382
xcg17206393 -0.348 -0.221    -0.312     -0.198
xcg02825373 -0.207 -0.174    -0.165     -0.138
xcg09255154 -0.199 -0.159    -0.497     -0.398
xcg19931348  0.118  0.146     0.140      0.172
xcg00340850 -0.061 -0.061    -0.057     -0.057
xcg27379887 -0.044 -0.036    -0.114     -0.093
xcg05139523  0.044  0.039     0.042      0.037
xcg19866811 -0.031 -0.013    -0.070     -0.031
xcg10925829 -0.030 -0.039    -0.116     -0.151
xcg04629133 -0.018 -0.006    -0.025     -0.008
xcg15560495  0.018  0.010     0.016      0.009
xcg02167438  0.017  0.005     0.012      0.004
xcg09572195 -0.013 -0.010    -0.012     -0.009
xcg09577651  0.011  0.011     0.014      0.014
xcg22676401 -0.011 -0.002    -0.007     -0.002
xcg00435238  0.010  0.009     0.010      0.008
xcg09210315 -0.009 -0.005    -0.016     -0.010
xcg10475341 -0.006 -0.002    -0.019     -0.006
xcg27144039 -0.006 -0.003    -0.017     -0.010
xcg20319775 -0.005 -0.004    -0.020     -0.016
-----------------------------------------------------------------
```

The second part of the above output lists the CpG selected along with their regression coefficients indicated by `gnet.scale`. By selecting important variables from a much larger set of candidate CpGs (9,125 CpGs), we identified 22 CpGs, none of which include the selected CpGs in Table 6.1. It is worthy noting that the candidate variables in `newx` are not among the 22 CpGs. Thus, although we found some CpGs showing associations with DNA methylation at `cg00493554` based on a smaller pool of 20 candidate CpGs, when searching through the 9,125 candidate CpGs, those selected CpGs from the smaller pool become much less important.

Table 6.1 Selected CpG sites for each of the methods discussed in this chapter. Selected CpGs common across all the approaches are in bold font.

Method	Selected CpGs	Reg. Coeff.	Method	Selected CpGs	Reg. Coeff.
Elastic Net	cg00156569	-0.39	Adaptive	cg00156569	-0.51
	cg00157199	-0.21	LASSO	cg00157199	-0.0072
	cg00188822	0.051		cg00226306	0.49
	cg00226306	0.47		**cg00254133**	-0.65
	cg00246386	-0.11		**cg00298324**	0.63
	cg00254133	-0.47			
	cg00298324	0.52			
	cg00309582	0.075			
SCAD	cg00226306	0.0091	*g*-prior	**cg00254133**	-0.92
	cg00254133	-1.18	(*g*=sample	**cg00298324**	0.65
	cg00298324	0.16	size)		
FLS	**cg00254133**	-0.92	Robust	**cg00254133**	-0.92
	cg00298324	0.65		**cg00298324**	0.65
Spike-and-slab	**cg00254133**	-1.09	Spike-and-slab	cg00156569	-0.46
(Coefficient)	**cg00298324**	0.30	(variance	cg00157199	-0.12
			component)	cg00226306	0.44
				cg00246386	-0.055
				cg00254133	-0.60
				cg00298324	0.62
				cg00309582	0.064

Non- and semi-parametric methods to select genetic and epigenetic factors

The methods discussed in Chapter 6 focus on selecting variables in parametric linear models. It is considered that genetic and epigenetic effects on the outcome are not linear and can be in any unknown form. Methods for feature selections in non-linear models have been developed [18, 116, 99, 126]. These methods are generally built upon splines or Taylor series expansions and may have difficulty in accommodating a large number of predictors and describing complex interaction effects. In genetic or epigenetic studies, the number of candidate predictors is relatively large due to biological uncertainty. Genes or epigenes (genes associated with epigenetic variants) do not necessarily function individually, rather, they work in concert with others to manifest a disease condition. In the area of machine learning, methods of variable selection in semi-parametric models constructed based on reproducing kernels have been discussed [121, 160, 153, 154], which are able to deal with complex interactions. Terry et al. [137] proposed a unified variable selection approach that has the ability to select variables from a mixture of continuous (e.g., DNA methylation) and ordinal variables (e.g., the number of alleles in a genotype, $0, 1,$ or 2). In this chapter, we will discuss two methods based on splines, Lin and Zhang and Sestelo et al. [99, 126], one approach based on Zadaraya-Watson kernel regression estimation proposed by Zambom and Akritas [153], and the methods by Zhang et al. [160] and Terry et al. [137], which utilize reproducing kernels.

In the following sections, we introduce each of the four approaches, followed by demonstrations of these methods in Section 7.4 using the epigenetic data discussed in Chapter 6 as well as simulated data.

7.1 VARIABLE SELECTION BASED ON SPLINES

Lin and Zhang [99] proposed an approach that has the ability to deal with non-linear models, named as the COmponent Selection and Smoothing Operator (COSSO). Compared to other penalty-based variable selection approaches, the penalty function in COSSO is the sum of component norms rather than the squared norms as in traditional smoothing spline methods. The penalized objective function in COSSO is defined as

$$
\begin{aligned}
\mathcal{O} &= \frac{1}{n} \sum_{i=1}^{n} \{Y_i - f(X_i)\}^2 + \tau_n^2 J(f) \\
J(f) &= \sum_{\alpha=1} p \|P^\alpha f\|,
\end{aligned} \tag{7.1}
$$

where p represents the number of orthogonal subspaces in a function space $\mathcal{F}$ and for the situation of main effects space p is the number of candidate variables. Parameter τ_n is a smoothing parameter to be tuned. The penalty $J(f)$ is the summation of reproducing kernel Hilbert space norms. It is a convex function and is a pseudo-norm. That is, for any $f, g \in \mathcal{F}$, $J(f) \geq 0$, $J(cf) = |c|J(f)$, $J(f + g) \leq J(f) + J(g)$, and for any nonconstant $f \in \mathcal{F}$, $J(f) > 0$. We have

$$
\sum_{\alpha=1}^{p} \|P^\alpha f\| \leq J^2(f) \leq p \sum_{\alpha=1}^{p} \|P^\alpha f\|^2.
$$

Lin and Zhang [99] show that COSSO exists due to the convexity of the penalized objective function. An equivalent form of the penalized objective function (7.1) is given as [99],

$$
\mathcal{O}_e = \frac{1}{n} \sum_{i=1}^{n} \{Y_i - f(X_i)\}^2 + \lambda_0 \sum_{\alpha=1}^{p} \theta_\alpha^{-1} \|P^\alpha f\|^2 + \lambda \sum_{\alpha=1}^{p} \theta_\alpha, \theta_\alpha \geq 0,
$$

where λ_0 can be any positive value and λ is a smoothing parameter that needs to be tuned. To tune λ, Lin and Zhang [99] propose to use 5-fold cross validation or generalized cross-validation to minimize risk.

The method of COSSO has been built into R and the R function cosso can be applied to different types of regression models such as regression to the mean, logistic regressions, quantile regressions, and Cox models.

Also using splines to describe non-linear associations, Sestelo et al. [126] developed an R package FWDselect for variable selection in an additive regression model,

$$Y_i \;=\; \beta_0 + m_1(X_{1i}) + \cdots + m_p(X_{pi}) + \epsilon_i, \qquad (7.2)$$

where $m_j, j = 1, \cdots, p$, are smooth and unknown functions, ϵ_i represents random errors with mean of zero, and $E(m_j(X_j)) = 0$ implying $E(Y) = \beta_0$. In Sestelo et al. [126], penalized regression splines are applied to estimate m_j.

The variable selection procedure has two parts. One part is to select the best combination of q variables for a given q. This is achieved by optimizing information criteria such as AIC or deviance and the selection of variables is through a forward stepwise-based procedure. The other part is to determine the optimal number of covariates to be included in the model, through hypothesis testing,

$$H_0(q) \quad : \quad \sum_{j=1}^{q} I_{\{m_j \neq 0\}} \leq q$$

$$H_1(q) \quad : \quad \sum_{j=1}^{q} I_{\{m_j \neq 0\}} > q, \qquad (7.3)$$

where I is an indicator function. The hypothesis testing is based on the contribution of additional $p - q$ variables on their ability to explain the variation in Y_i. An empirical p-value is estimated based on wild bootstrap samples for the purpose of making a decision on the hypothesis testing. To select variables, two functions in the package FWDselect are needed, selection and test. Function test is used to determine the best q, and function selection is then applied to select the best combination of q variables. Alternatively, function qselection can be applied. For each sub-model, this function calculates the information criterion, e.g., AIC or deviance, based on which the best model can then be determined.

7.2 OVERVIEW OF THE ANOVA-BASED APPROACH

The approach proposed by Zambom and Akritas [153], denoted as the ANOVA-based approach throughout this section, applies kernel regression to estimate any non-linear associations.

Define a regression model in the following structure,

$$Y_i = m_1(\boldsymbol{X}_{1i}) + m_2(X_{2i}) + \epsilon_i, \ i = 1, \cdots, N, \tag{7.4}$$

where $m_1(\boldsymbol{X}_{1i})$ represents an unknown function describing the association between Y_i and $p - 1$ dimensional vector $\boldsymbol{X}_{1i}$, $m_2(X_{2i})$ is the other component in the regression with $m_2(\cdot)$ unknown and $X2$ a univariate variable, and ϵ_i is the random error such that $\epsilon_i \sim N(0, \sigma^2)$. The null hypothesis is

$$H_0 \ : \ m_2(X_{2i}) = 0,$$

that is, $E(\boldsymbol{Y}|\boldsymbol{X_1}, X_2) = m_1(\boldsymbol{X_1})$ under H_0.

To deal with high dimensional data, Zambom and Akritas [153] adopted the idea of Akritas and Papadatos [2], which performs hypothesis testing in high-dimensional one-way analysis of variance analyses (ANOVA). For regressions, however, $m_1(\boldsymbol{X}_{1i})$ is unknown and for each specific measure of a variable in $\boldsymbol{X_1}$ (noted as "cell" in Zambom and Akritas [153]), in general there will be only one observation of $\boldsymbol{Y}$. To utilize the technique developed for high-dimensional one-way ANOVA, Zambom and Akritas [153] estimated $m_1(\boldsymbol{X}_{1i})$ using Zadaraya-Watson kernel regression estimation,

$$\hat{m}_1(\boldsymbol{X}_{1i}) = \sum_{j=1}^{N} \Big(\frac{K_{H_n}(\boldsymbol{X}_{1i} - \boldsymbol{X}_{1j})}{\sum_{l=1}^{N} K_{H_n}(\boldsymbol{X}_{1i} - \boldsymbol{X}_{1l})} \Big) Y_j, i = 1, \cdots, N,$$

where $K_{H_n}(\boldsymbol{x}) = |H_n|^{-1} K(H_n^{-1}\boldsymbol{x})$, with $K(\cdot)$ being a kernel with a bounded width, and H_n is a positive definite symmetric matrix with dimension $(d - 1) \times (d - 1)$ noted as a bandwidth matrix. To solve the single observation per "cell' problem, Zambom and Akritas [153] proposed to augment the data by artificially grouping X_{2i}'s into groups. The augmentation is on X_{2i} due to the construction of H_0. With these two problems solved, to perform variable selections, a test statistic defined as $n^{1/2}(MST - MSE)$ is used to carry out a testing on H_0 with MST denoting mean squares for treatment and MSE for mean squares of error. Clearly, this approach does not have the ability to select variables if the number of candidate variables is larger than the sample size.

Based on p-values inferred from the test statistic, Zambom and Akritas proposed backward and forward approaches to select variables associated with dependent variable Y, which are available in the R package `NonModelCheck` [154].

The R package

The R function `npvarselec` in the package `NonModelCheck` [154] implements the methods proposed by Zambom and Akritas [153]. In total, three selection algorithms are available, `backward`, `forward`, and `forward2`. The `backward` selection procedure is introduced in Zambom and Akritas [153]. It starts from a model with all the variables included. Then at each step, the algorithm eliminates the least significant variable if its p-value is larger than a cut off value after correcting for false discovery rate (FDR) applied to dependent tests [8]. Correcting for multiple testing assuming dependent tests is under the consideration of potentially joint activities among genes. The `forward` selection approach, on the other hand, starts from a null model. At each step, it temporarily adds in a variable with the smallest p-value, taking into account the variables already in the model. This candidate variable will be retained if each variable in the new model is statistically significant after controlling for FDR. The `forward2` approach is a slight extension of `forward`. In this case, a variable is added to the model if this new variable results in the most significant new model, even though this variable itself is not the most significant. For the purpose of comparing between different methods, we hold up our demonstration of the package until a later section, Section 7.4.

7.3 VARIABLE SELECTION BUILT UPON REPRODUCING KERNELS

Zhang et al. [160] proposed a variable selection approach targeted at variables associated with a dependent variable and the association can be in a complex form. In their approach, they allow a set of variables, Z, not to participate in the selection and those variables will be kept in the final model. Another set of variables, X, are candidate variables and any variables in X not associated with the response variable Y will be excluded. To model the association of Y with Z and X, we consider the following setting to evaluate their effects,

$$Y = Z\beta + h(X) + \epsilon, \tag{7.5}$$

where Y is a vector of length N, and linear associations are assumed between Y and Z. The association of Y and X is defined by function $h(\cdot)$, which is possibly complex and non-linear. The assumption on random error ϵ is the same as before, i.e., $\epsilon \sim N(0, \sigma^2 I)$ with I being an identity matrix. All variables in Z will be kept in the model and β describes additive linear effects of Z. Assume in total p variables in X and not all variables in X are important. Unimportant ones in X need to be identified and excluded. Function $h(\cdot)$ is described through a kernel function $K(\cdot, \cdot)$. By Mercer's theorem [109, 23], under some regularity conditions, the kernel function $K(\cdot, \cdot)$ specifies a unique function space $\mathcal{H}$ spanned by a particular set of orthogonal basis functions. The orthogonality is defined with respect to L_2 norm. Following Mercer's theorem, any function $h(\cdot)$ in the function space $\mathcal{H}$ can be represented as a linear combination of reproducing kernels [23, 55], $h(X_i) = \sum_{l=1}^{N} K(X_i, X_l)\alpha_l = K_i'\alpha$, where $\alpha = (\alpha_l, l = 1, \cdots, N)'$ is a vector of unknown parameters and K_i' is the ith row of kernel matrix K. This definition of $h(\cdot)$ introduces a potential to capture complex interactions between variables in X via the kernel function K and has the ability to handle a large number of variables. In Zhang et al. [160], a Gaussian kernel, $K(\rho)$, is suggested with entry k_{il} defined as

$$k_{il}(\rho) = \exp\left\{-\sum_j ||X_{ij} - X_{lj}||^2/\rho\right\},$$

with $i, l = 1, \cdots, N$, $j = 1, \cdots, p$, where X_{ij} is the measure of variable j of subject i and ρ is a regularization parameter.

To select variables from X, an indicator variable $\delta = \{\delta_j | j = 1 \cdots, p\}$ is further included into the kernel matrix with $\delta_j = 1$ indicating the inclusion of variable j and 0 otherwise. With the involvement of δ_j, we denote the kernel matrix as $K(\rho, \delta)$ with its (i, l)-th entry defined as

$$k_{il}(\rho, \delta) = \exp\left\{-\sum_j ||(X_{ij} - X_{lj})\delta_j||^2/\rho\right\}.$$

If variable j is excluded, then it will not appear in any entry of the kernel matrix. Under this setting, Zhang et al. [160] developed a fully Bayesian approach to infer parameters β, σ^2 and δ. For the regularization parameter ρ, we see that different values of ρ with different sets of selected variables can result in the same $\boldsymbol{K}(\rho)$ and consequently the same likelihood. Thus, it is suggested to fix ρ at $\rho = \rho_0$ with ρ_0 being the value estimated at the full model with all variable included. This is under the consideration that unimportant variables do not significantly contribute to the joint effect of $\boldsymbol{X}$. In addition, through sensitivity analyses on variable selection results with respect to different choices of ρ_0, Zhang et al. [160] suggests take $\rho_0 = 1$.

Each candidate variable is selected based on its posterior probability of being included in the model. A posterior probability of a variable is calculated as the percentage of times that the variable is selected among a certain number of uncorrelated Markov Chain Monte Carlo samples. To make a decision, the concept of a scree plot is applied to the posterior probabilities. Scree plots are often used in principal component analysis to determine the number of components, where a sharp decrease in eigenvalues indicates less importance for the rest of the components. Analogously, a sharp decrease in sorted posterior probabilities indicates less importance of the remaining variables. Variables identified by this rule are treated as the most important variables.

This method is not available as an R package, but an R program (VarSelKernelFunctionsContinuous.R) that implements the method can be downloaded from `https://www.memphis.edu/sph/contact/faculty_profiles/zhang.php`.

Dealing with a mixture of continuous and ordinal variables

The approaches discussed so far are designed for continuous variables. In genetic and epigenetic studies on various health conditions, potentially joint activities of single nucleotide polymorphisms (SNPs) and DNA methylation to manifest a disease status have drawn investigators' attention [161, 133, 86]. Methods with the ability to select informative SNPs and DNA methylation sites under the consideration of their possibly joint work are of great interest. To fit this need, extended from the method by Zhang et al. [160], Terry et al. [137] proposed an approach to select variables that are a mixture of ordinal and continuous variables. They standardize both the ordinal and continuous variables for the purpose to diminish the scale difference between these two types variables,

$$g_l^* = \frac{g_l - \overline{g_l}}{s_l}. \tag{7.6}$$

The Gaussian kernel is applied to the standardized variables as in Zhang et al. [160] to describe complex associations between a set of candidate variables and an outcome of interest,

$$K(g_{v1}^*, g_{v2}^*) = \exp\{-\|g_{v1}^* - g_{v2}^*\|^2/\rho_0\}. \tag{7.7}$$

By standardizing, all variables are measured on the same scale. The functionality of the Gaussian kernel is invariant to this transformation in that it still evaluates the distance between subjects. Standardizing all variables has some drawback for ordinal variables of yielding only a few different values say $(0,1,2)$, but based on previous studies, such transformations do not seem to bias inferences [168, 81].

For the selection of ρ_0, note that $Var(g_l^*) \to 1$ as $n \to \infty$. This leads to $Var(g_{l1}^* - g_{l2}^*) \to 2$ assuming no correlations between the predictors. The stabilization in the predictors due to standardization suggests setting ρ_0 as $\rho_0 = 2$. The effectiveness and robustness of such a selection seems to be supported by various simulations discussed in Terry et al. [137].

To apply the method in [137], the data are standardized first, and then we can apply the same program in Zhang et al. [160] with the regularization parameter set at 2 or use the approaches in the package `npvarselec`. For variables that are a mixture of nominal and continuous variables, it seems that no specific non-parametric variable selection approaches are available to deal with this type of data.

7.4 EXAMPLES

In this section, we demonstrate the four approaches introduced in this chapter by applying them to the epigenetic data discussed in Chapter 6, where variables are selected based on their linear associations with the dependent variable, as well as simulated data sets where we know the underlying truth.

7.4.1 Selecting important epigenetic factors

Recall, the following codes are used to extract the data and perform screening to identify CpGs potentially associated with a marker of interest, i.e., DNA methylation at CpG site `cg00493554`.

```
> library(MASS)
> library(IlluminaHumanMethylation450kmanifest)
> library(IlluminaHumanMethylation450kanno.ilmn12.hg19)
```

```
> library(IlluminaHumanMethylationEPICanno.ilm10b4.hg19)
> library(IlluminaHumanMethylationEPICmanifest)
> library(prettyunits)
> library(bumphunter)
> library(DMRcate)
> data(dmrcatedata)
> DNAm = logit2(myBetas)
> dim(DNAm)
[1] 10042    76
> DNAmNoProbSNPs = rmSNPandCH(DNAm, dist = 10, mafcut = 0.05)
> dim(DNAmNoProbSNPs)
[1] 9126    76
> CpGs = rownames(DNAmNoProbSNPs)
> ID = factor(sub("-.*", "", colnames(DNAm)))
> Status = factor(sub(".*-", "", colnames(DNAm)))
> library(ttScreening)
> runs = ttScreening(y = DNAmNoProbSNPs,
formula = ~Status,data=Status,
+ imp.var = 1, sva.method = "two-step",
+ B.values = FALSE, iterations = 100, cv.cutoff = 50,
+ n.sv = NULL, train.alpha = 0.05, test.alpha = 0.05,
+ FDR.alpha = 0.05, Bon.alpha = 0.05, percent = (2/3),
+ linear = "ls", vfilter = NULL, B = 5,
+ numSVmethod = "be", rowname = NULL,maxit = 20)
Number of significant surrogate variables is:   22
[1] "9 surrogate variables used in the analysis"
> out = runs$TT.output
> dim(out)
[1] 667   24
> maxFreq = max(out[,2])
> maxFreq
[1] 100
> rowMaxFreq = as.numeric(as.character((out[which(out[,2]
+ ==maxFreq),1]))))
> y = t(DNAmNoProbSNPs[rowMaxFreq,])
> rownames(y) = CpGs[rowMaxFreq]
> rownames(y)
[1] "cg00493554"
> x = t(DNAmNoProbSNPs[-rowMaxFreq,])
> colnames(x) = CpGs[-rowMaxFreq]
```

Results from cosso We start from the method by Lin and Zhang, implemented in the R package cosso. Since the variable selection approach by Lin and Zhang [99] requires the number of variables smaller than the sample size, we use the first 20 variables to demonstrate. These 20 variables are the same as those discussed in Chapter 6. Function `cosso` is used to select the variables. Our goal is to identify DNA methylation sites such that DNA methylation at those sites is associated with DNA methylation at `cg00493554`. It is reasonable to use regression to the mean with a Gaussian distribution specified for random errors. We thus set `family="Gaussian"` in function `cosso`. We implement the following R codes to select CpGs.

```
> y = as.vector(y)
> results = cosso(newx,y,family = "Gaussian")
> colnames(results$tune$L2norm) = colnames(newx)
> norms = results$tune$L2norm[(ncol(newx)+2),]
```

Along with other outputs, the object `results` provides L_2-norms evaluating the contribution of each variable, extracted by `results$tune$L2norm`. In `results$tune$L2norm`, all the L_2-norms calculated in the tuning process are included. In the above, we use the last set of L_2-norms to choose important variables using the scree plot. The R codes are below,

```
> names(norms) = colnames(results$tune$L2norm)
> sortedNorm = sort(norms,decreasing=TRUE)
> plot(sortedNorm,pch = 19,cex = 0.5,ylab = "Sorted L2-norm")
> lines(sortedNorm)
> names(sortedNorm[1:3])
[1] "cg00171166" "cg00168514" "cg00309582"
```

Based on the scree plot using the sorted L_2-norms `sortedNorm`, three CpG sites are selected, cg00171166, cg00168514 and cg00309582.

Results from FWDselect We first use the function `test` to determine the number of variables to be included in the model.

```
> selectq = test(newx, y, method = "gam", family = "gaussian",
+ nboot = 50, speedup = TRUE, bootseed = 12345, cluster = TRUE)
[1] "Processing IC bootstrap for H_0 ( 1 )..."
**************************************
  Hypothesis Statistic pvalue     Decision
1    H_0 (1)      15.86    0.06 Not Rejected
> selectq$nvar
[1] 1
```

In this function, generalized additive models `gam` are used to assess the associations and 50 bootstrap samples are used to determine q. We set `speedup=TRUE` to increase the computing efficiency. In this case, when testing the hypothesis, the alternative models consider one more variable than the number of variables in the null hypothesis. Although this will speed up the testing process, wrong decisions can be made due to the limited assessment on models. Setting `cluster=TRUE` is to indicate that the testing procedure is parallalized to increase computing efficiency. Applying `test` to the 20 CpG sites, $q = 1$ is determined by calling `results$nvar`.

Next, given $q = 1$, we use `selection` to select the best variable from the 20 candidate variables.

```
> q = selectq$nvar
> results = selection(newx, y, q = q, prevar = NULL,
+ criterion = "deviance", method = "gam", family = "gaussian",
+ nmodels = 1, nfolds = 5, cluster = TRUE)
> results
****************************************************
Best subset of size q = 1 : cg00254133
Information Criterion Value - deviance : 8.511648
****************************************************
```

The information criterion `deviance` calculated using 5-fold cross validation (`nfolds=5`) is used to select the best models. Through this approach, CpG site `cg00254133` is selected, which is not in any of the three CpG sites selected from `cosso`.

Alternatively, with similar settings, we can use `qselection` to identify a set of best models with different sizes with each determined by a pre-specified information criterion, e.g., deviance. Compared to the method in `cosso`, the method in `FWDselect` does not have a good control on the number of knots used in the regression splines, which limits the number of variables that are allowed in a model. In the following program, the maximum number of variables is set at 8. After comparing best models from size 1 to size 8, a model with one CpG site, `cg00254133`, is selected as the best model, indicated by ∗ corresponding to the smallest deviance. This is the same result as that from using `test` and `selection` together.

```
> results = qselection(newx, y, qvector = c(1:8),
+ criterion = "deviance", method = "gam", family = "gaussian",
+ nfolds = 5, cluster = TRUE)
```

```
[1] "Selecting subset of size 1 ..."
[1] "Selecting subset of size 2 ..."
[1] "Selecting subset of size 3 ..."
[1] "Selecting subset of size 4 ..."
[1] "Selecting subset of size 5 ..."
[1] "Selecting subset of size 6 ..."
[1] "Selecting subset of size 7 ..."
[1] "Selecting subset of size 8 ..."
> results
  q deviance
1 1    8.512
2 2    8.793
3 3    12.86
4 4   39.136
5 5   67.228                                        cg00254133,
6 6   23.415                          cg00254133, cg00260888,
7 7   91.933              cg00254133, cg00260888, cg00105628,
8 8   27.257 cg00254133, cg00260888, cg00105628, cg00303672,
>
  q deviance                                        selection
1 1    8.512                                        cg00254133 *
2 2    8.793                          cg00254133, cg00168514
3 3    12.86              cg00254133, cg00168514, cg00105628
4 4   39.136 cg00254133, cg00168514, cg00105628, cg00303672
5 5   67.228 cg00260888, cg00105628, cg00303672, cg00273813
6 6   23.415 cg00105628, cg00303672, cg00273813, cg00199007
7 7   91.933 cg00303672, cg00273813, cg00156569, cg00159400
8 8   27.257 cg00273813, cg00298324, cg00159400, cg00188822
```

Results from the ANOVA-based approach As in the method COSSO, the variable selection approach by Zambom and Akritas [153] does not allow the number of variables larger than the sample size. In the following, we use the same 20 variables.

```
> newx = x[,1:20]
> library(NonpModelCheck)
> out = npvarselec(newx,as.vector(y),method = "backward",
+          degree.pol = 1,kernel.type = "trun.normal",
+             bandwidth = "CV", dim.red = c(1, 10))
----------------------------------------------------------------
Iter. | Variables in the model
1     | 1 2 3 4 5 6 7 8 9 10 11 12 13 14 15 16 17 18 19 20
2     | 1 2 3 4 5 6 7 8 10 11 12 13 14 15 16 17 18 19 20
```

```
3      | 1 2 3 4 5 6 7 8 10 11 12 13 14 15 16 17 18 19
4      | 1 2 3 4 5 6 7 8 10 11 12 13 14 16 17 18 19
5      | 1 2 3 4 5 6 7 8 11 12 13 14 16 17 18 19
6      | 1 2 3 4 5 6 7 8 11 13 14 16 17 18 19
7      | 1 2 3 4 5 6 7 8 11 13 14 16 17 18
8      | 1 2 4 5 6 7 8 11 13 14 16 17 18
9      | 2 4 5 6 7 8 11 13 14 16 17 18
10     | 2 4 5 6 7 8 11 13 14 16 18
11     | 4 5 6 7 8 11 13 14 16 18
12     | 4 5 6 7 8 11 13 14 18
13     | 4 5 6 7 8 13 14 18
14     | 4 6 7 8 13 14 18
15     | 4 6 8 13 14 18
16     | 4 6 13 14 18
17     | 4 13 14 18
18     | 13 14 18
----------------------------------------------------------------
> out
Number of Covariates Selected: 3
Covariate(s) Selected:
--------------------------
Covariate Index  |  p-value
           13    |  0.0053
           14    |  1e-10
           18    |  1.3e-09
--------------------------
> # selected variables
> colnames(x)[out[[1]]]
[1] "cg00246386" "cg00254133" "cg00298324"
```

The last two selected CpG sites, cg00254133 and cg00298324, are also selected by all the methods noted in Table 6.1 in Chapter 6. Applying the two forward selection approaches, forward and forward2, different variables are selected. In particular, with forward, only one CpG site, cg00105628, is selected, and with forward2, cg00105628, cg00246386, and cg00205332 are selected. However, none of these CpGs are selected when the backward selection approach was applied.

In the codes above, kernel.type specifies the kernels used to fit local polynomial regression. In the above, trun.normal kernel is selected. It is a Gaussian kernel truncated between -3 and 3. The argument bandwidth is the bandwidth in the kernel and is determined based on cross validation by setting bandwidth="CV". The argument degree.pol specifies the order of local polynomials used to fit the unknown associations. After increasing the degree, the results overall show no difference with `degree.pol=1`. The last argument `dim.red` is for dimension reduction aiming to improve computing efficiency. `dim.red` is a vector with the first element being 1 denoting sliced inverse regression (SIR) and 2 for supervised principal components (SPC). The second element in the vector gives the number of slices (if SIR), or the number of principal components (if SPC). If setting `dim.red=0`, then no dimension reduction is performed. We suggest not setting it to zero but choosing between SIR and SPC for the purpose of increasing the computing speed.

Results from the reproducing kernel-based approach   We next apply the method proposed by Zhang et al. [160] to this epigenetic data. We use `source` to call the function. The dependent variable Y is the same as before, DNA methylation at CpG site cg00493554. Since we do not have any covariates assumed to be linearly associated with the response variable, the design matrix for the linear part is a vector of 1's, and we set `numCov=0`. The regularization parameter ρ_0 is set at 1 following suggestions of Zhang et al. [160]. Arguments a0 and b0 are the hyper parameters in the prior distribution of the random error variance parameter σ^2 in the regression model. These two parameters are chosen as being non-informative and set at 0.001. Two chains of Markov Chain Monte Carlo simulations are run, which can be used to assess the potential of convergence.

```
> source("VarSelKernelFunctionsContinuous.R")
> Y = as.vector(y)
> X = rep(1,length(Y))
> Candid = newx
> VarSelResults = VarSel(Y, X, Candid, rho = 1, numCov = 0,
+ numItr = 100, numChain = 2, a0 = 0.001,b0 = 0.001)
```

The method of Zhang et al. identifies three CpG sites, cg00105628, cg00254133, cg00298324, such that DNA methylation at these three sites is possibly associated with DNA methylation at cg00493554. The sorted estimated probabilities of variable selection in ascending order are stored in `orderProp`, as indicated in the codes below. The estimated probability of selecting the top three CpG sites is 1, which is substantially larger than the fourth CpG site to be selected, implying the importance of these three CpG sites. These three selected CpGs overlap with the ones selected using the three approaches in `npvarselec`.

```
> colnames(Candid)[VarSelResults$VarSelect]
```

```
[1] "cg00105628" "cg00254133" "cg00298324"
> VarSelResults$orderProp
      Variable Index Post. Prob
  [1,]               2  1.0000000
  [2,]              14  1.0000000
  [3,]              18  1.0000000
  [4,]               5  0.1333333
  [5,]               1  0.0000000
  [6,]               3  0.0000000
 ......
```

In this example, the data are continuous. In the situation of a mixture of continuous and ordinal data, the method by Terry et al. [137] can be applied by first standardizing the data and then applying the program VarSelKernelFunctionsContinuous.R. Compared with the methods in cosso, FWDselect and npvarselec, the methods by Zhang et al. and Terry et al. [160, 137] have the potential to deal with large numbers of candidate variables such that the number of independent variables is larger than the sample size.

7.4.2 Selecting variables with known underlying truth

To have a better understanding of the methods in terms of their performance in different situations, we simulated a data set based on the model below,

$$E(Y_i|\boldsymbol{X}_i) = 3\cos(X_{i1} \times X_{i2}) + 2X_{i3}, \ i = 1, \cdots, N,$$

where $\boldsymbol{X}_i$ is a vector of length $p = 12$ representing a set of candidate variables and $N = 100$. Among the 12 candidate variables, the first three are important variables such that Y is non-linearly associated with the interaction of X_1 and X_2, and linearly associated with X_3. A simple program to generate the data is below, in which the candidate variables are generated from uniform distributions each with different lower and upper bounds.

```
> set.seed(seed = 1000)
> n = 100
> numCov = 0
> numX = 12
> #generate candidate variables
> x = matrix(rep(0,numX*n),ncol = numX)
```

```
> for (i in 1:numX)
+ {
+ x[,i] = runif(n,0.0001, numX/(2*i))
+ }
> # non-linear interaction
> xInter = 3*cos(x[,1]*x[,2])
> delta = c(0,0,1,rep(0,numX-3))
> xEff = rep(2,length = numX)*delta
> #generate observations of y
> mu = x%*%xEff+xInter
> Var = diag(0.25, n, n)
> y = rmnorm(1,mu,Var)
> y = t(y)
> data = cbind(t(y),x)
```

We apply the methods discussed so far to the simulated data, starting from the method in cosso. The program is comparable to the program discussed earlier for the epigenetic data. Again, we use a scree plot to assist our decision on the number of important variables by plotting the sorted L_2-norms. For the purpose of illustration, we include the scree plot in Figure 7.1. The plot suggests three important variables and the three correct variables 1, 2, and 3 are selected by checking the names of sortedNorm.

```
> Y = data[,1]
> X = rep(1,length(Y))
> Candid = as.matrix(data[,-1])
> results = cosso(Candid,Y,family = "Gaussian")
> colnames(results$tune$L2norm) = seq(1,ncol(Candid))
> norms = results$tune$L2norm[(ncol(Candid)+2),]
> names(norms) = colnames(results$tune$L2norm)
> sortedNorm = sort(norms,decreasing = TRUE)
> plot(sortedNorm,pch = 19,cex = 0.5,ylab = "Sorted L2-Norm")
> lines(sortedNorm)
> names(sortedNorm[1:3])
[1] "2" "1" "3"
```

The two approaches in the package FWDselect give different results. Similar programs as in the real data applications are applied and thus they are not included here. With selection paired with test, variables 1 and 3 are selected, while with qselection, including variables 1 to 3 plus two additional variables, variables 8 and 11, give the smallest deviance. Accompanied by the real data application, the methods in FWDselect seem to not perform well in general.

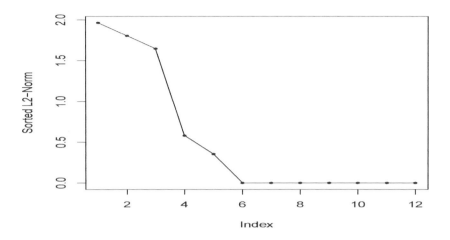

Figure 7.1   The scree plot of L_2-norms for the simulated data.

We next apply the backward selection method in `npvarselec` with the following codes. Again, we omit the program here to avoid duplications. With the backward selection method, variables 3 and 8 are selected. The other two approaches, `forward` and `forward2`, give the same results. The true variables 1 and 2 are not selected by any of the three approaches. It is unclear whether this is due to the Zadaraya-Watson kernel regression estimation of $m_1(\boldsymbol{X}_{1i})$ in the model of (7.4).

Finally, we apply Zhang et al. method to the same data,

```
> Y = data[,1]
> X = rep(1,length(Y))
> Candid = as.matrix(data[,-1])
> VarSelResults = VarSel(Y, X, Candid, rho = 1, numCov = 0,
+ numItr = 500, numChain = 1, a0 = 0.001,b0 = 0.001,sigB2 = 10)
> VarSelResults$VarSelect
[1] 1 2 3
>
> VarSelResults$orderProp
        Variable Index Post. Prob
 [1,]                1    1.0000
 [2,]                2    1.0000
 [3,]                3    1.0000
 [4,]               11    0.0848
 [5,]               12    0.0848
 . . . . . .
```

The first three variables, which are the truly important variables, are selected by the Zhang et al. method with probability 1. For this simulated data, the method in Zhang et al. [160] outperforms the three approaches in the package npvarselec.

We further apply Zhang et al. method to a situation such that the number of variables is larger than the sample size. Similar simulation scenarios are applied except that the number of candidate variables is increased to 100 with a sample size of 100 and the candidate variables are generated from uniform distributions with lower and upper bounds specified as the following,

```
> for (i in 1:numX)
+ {
+ x[,i] = runif(n,0.0001, numX/(20*i))
+ }
```

Including the newly simulated data in the data matrix as done earlier and running the function VarSel with similar settings as before,

```
> VarSelResults = VarSel(Y, X, Candid, rho = 1, numCov = 0,
+ numItr = 100, numChain = 1, a0 = 0.001,b0 = 0.001,sigB2 = 10)
```

The three variables are all correctly selected with a posterior selection probability of 1 and are the only three variables selected based on the scree plot noted earlier, although the posterior probabilities of 30 other variables are all ≥ 0.5.

```
> VarSelResults$VarSelect
[1] 1 2 3
> VarSelResults$orderProp
```

	Variable Index	Post. Prob
[1,]	1	1.00000
[2,]	2	1.00000
[3,]	3	1.00000
[4,]	50	0.68750
[5,]	63	0.65625
[6,]	92	0.65625
[7,]	48	0.62500
[8,]	100	0.59375
[9,]	28	0.56250
. . . . . .		
[15,]	99	0.56250
[16,]	29	0.53125
. . . . . .		
[21,]	69	0.53125

```
[22,]              34    0.50000
......
[33,]              98    0.50000
[34,]              21    0.46875
......
[98,]               5    0.03125
[99,]               4    0.00000
[100,]              6    0.00000
```

Among all the four approaches discussed in this chapter, in the DNA methylation example, findings from the method by Zhang et al. [160] agree with most of the other three approaches; in the example based on simulated data, the methods in cosso and Zhang et al. [160] perform the best. The advantage of Zhang et al., as well as Terry et al. [137], exists in their ability to select variables such that the number of candidate variables is larger than the sample size.

## CHAPTER 8

Network construction and analyses

In this chapter, we focus on Gaussian undirected networks or graphs and Bayesian (directed acyclic) networks as it is closely related to statistical modeling. In this type of graphic modeling, distributions or conditional distributions of data are assumed to be normal, multivariate normal for Gaussian undirected networks and univariate normal for Bayesian networks. In the following, "networks" and "graphs" are used interchangeably.

8.1 UNDIRECTED NETWORKS

Assume in total we have p variables (or genes) with measurement observed from n subjects and we would like to infer the connections among these variables. Let $\boldsymbol{X} = \{X_1, \cdots, X_k, \cdots, X_p\}$ with X_k denoting a variable at node k such that $\boldsymbol{X}$ follows a multivariate Gaussian distribution with covariance matrix Σ. We further denote by G_Σ the underlying true graph. Different approaches have been taken to infer G_Σ. Some methods are designed based on conditional covariances or partial covariance [146] by testing whether off-diagonal entries of an empirical partial covariance matrix are zero [147, 17]. This is under the assumption that an edge in G_Σ between node k_1 and k_2 exists if and only if the conditional covariance of X_{k1} and X_{k2} given all the other variables is non-zero. Precision matrix, $\Omega = \Sigma^{-1}$, has been commonly used to infer undirected graphs [152, 144]. An entry, (k_1, k_2), in the off-diagonal of Ω is non-zero if and only if there is an edge between nodes k_1 and k_2 in graph G_Σ. Other approaches utilize regressions to identify edges by regressing one node on the remaining nodes [108, 124], since a regression coefficient is non-zero if and only if there is an edge connecting the two nodes. In this section, we focus on the method proposed by Giraud et al. [58].

8.1.1 The two-stage graphs selection method

Giraud et al. in 2012 [58] proposed an approach that is computationally efficient with the ability to handle a large number of nodes. It is a two-stage approach. The basic idea is to first build a set of data-driven candidate graphs, and then select the best graph by minimizing the loss defined in Giraud [57]. Giraud et al. [58] proposed four ways to construct candidate graphs and suggested either taking a union of all the graphs from the four approaches or focusing on graphs from one approach at the stage of best graph selection.

One of the four approaches for graph construction is based on $0 - 1$ conditional independence graph. A $0 - 1$ conditional independence graph, G_{01}, is defined based on correlations and conditional correlations. That is, there is an edge between two nodes k_1 and k_2 if and only if the correlation between X_{k_1} and X_{k_2} is non-zero and the conditional correlation is non-zero between the two nodes given another node. Each $\hat{G}_{01}$ is constructed based on a likelihood ratio test at a level of α, the larger the α the more edges in $\hat{G}_{01}$. A set of graphs are selected such that α is small enough to ensure that the degree of each graph is less than a pre-specified integer $D < n - 2$. Often, $\hat{G}_{01}$ does not coincide with a Gaussian graph, G_Σ, although it is closely connected to G_Σ in some cases.

The second approach discussed in Giraud et al. constructs graphs based on the LASSO penalty [108] using the LARS-LASSO algorithm proposed by Efron et al. [35]. This approach estimates regression coefficients by

$$\hat{\theta}^\lambda = \text{argmin}\{\|\boldsymbol{X} - \boldsymbol{X}\boldsymbol{\theta}'\|^2_{n \times p} + \lambda\|\boldsymbol{\theta}'\|_1 : \boldsymbol{\theta}' \in \Theta\}, \qquad (8.1)$$

where Θ is a $p \times p$ matrix of regression coefficients and $\|\boldsymbol{\theta}'\|_1 = \sum_{k_1 \neq k_2} |\boldsymbol{\theta}'_{k_1,k_2}|$. The graph is then defined based on the status of regression coefficients. There is an edge between two nodes k_1 and k_2 if both $\hat{\theta}^\lambda_{k_1,k_2}$ and $\hat{\theta}^\lambda_{k_2,k_1}$ are non-zeros. The size of the graph tends to increase as λ increases. This family of graphs are chosen such that λ are large enough to ensure that the degree of a graph is at most D.

The third method is similar to the LASSO-based approach and the only difference is the utilization of adaptive LASSO. In this modified approach, the penalty in (8.1) is replaced by $\|\boldsymbol{\theta}'/\hat{\boldsymbol{\theta}}^{\text{init}}\|_1$ with $\hat{\boldsymbol{\theta}}^{\text{init}}$ being the initial estimator of $\boldsymbol{\theta}$, e.g., the exponential weights (EW) initial estimator [24] as suggested by Giraud et al. [58]. In this approach, nodes k_1 and k_2 are connected if $\hat{\theta}_{k_1,k_2}^{EW,\lambda}$ or $\hat{\theta}_{k_2,k_1}^{EW,\lambda}$ are non-zeros, different from the criterion used in the second approach.

The last approach in Giraud et al. [58] identifies graphs based on a good estimator of the true neighborhood of node k_1. It constructs two nested graphs, $\hat{G}_{and}$ and $\hat{G}_{or}$. There is an edge in graph $\hat{G}_{and}$ between nodes k_1 and k_2 if node k_1 is in the neighborhood of node k_2 and node k_2 is in the neighborhood of node k_1. For graph $\hat{G}_{or}$, neighborhood inclusion is required for one of the two nodes. Under this setting, $\hat{G}_{and}$ is nested with $\hat{G}_{or}$. The whole family of graphs under this setting is a collection of all the graphs between $\hat{G}_{and}$ and $\hat{G}_{or}$. Since the final graph will be selected through a search across all these possible graphs, this approach is denoted by Giraud et al. [58] as the quasi exhaustive search approach and can be time consuming.

We can take the union of all the graphs from the four approaches or a subset of these approach as a pool of candidate graphs. To select the best graph among the candidate graphs, define

$$\text{Crit}(G) \;=\; \sum_{k=1}^{p}\left[\|\boldsymbol{X}_k - \boldsymbol{X}[\hat{\theta}]_k\|^2\Big(1 + \frac{\text{pen}[d_k(G)]}{n - d_k(G)}\Big)\right],$$

where the penalty "pen" is a function of number of nodes and the degree of node k in the graph G, $d_k(G)$. This criterion is to identify a graph such that it minimizes the sum of penalized norms at each node on its distance from an estimated graph.

8.1.2 The GGMselect package and gene expression examples

The approach by Giraud et al. [58] has been built into a package GGMselect. In this package, the previously discussed four methods are implemented in two functions for graph selection, selectFast and selectQE. In the function selectFast, the method to construct $0 - 1$ conditional independence graphs (noted as C01), the LARS-LASSO approach (noted as LA), and the adaptive LASSO-based approach (noted as EW) are included, and in selectQE the quasi exhaustive search approach.

We demonstrate the two functions using a gene expression data set built into the package depthTools. Gene expressions were measured using the Affymetrix technology. Pre-processed expressions of 100 genes in 25 normal samples and 25 tumoral prostate samples are included in `depthTools`. These 50 samples are a random sample from the original study which has 50 normal and 52 tumoral prostate samples [129]. The raw data were preprocessed by taking thresholds at 10 and 16,000 units, excluding genes with expression variations less than 5-fold relatively or less than 500 units absolutely between samples, applying a base 10 log-transformation, and standardizing each experiment to zero mean and unit variance across the genes. The 100 genes available in `depthTools` and studied in Dudoit et al. [30] were the most variable genes in expressions such that ratios of between-group to within-group sum of squares in expression of genes were the largest among all the genes in the data [30].

We first call in the data set, named `prostate`. It has 50 rows and 100 columns. The first 100 columns are the expression levels of the 100 genes and the last column is the tumor status of each sample. We then give a generic name to each gene and name the last column as `TumorStatus`.

```
> # Call in the gene expression data of prostate samples
> library(depthTools)
> data(prostate)
> dim(prostate)
[1]   50 101
> names = NULL
> for (i in 1: (ncol(prostate)-1))
+ {
+ names = c(names, paste("g",i,sep = ""))
+ }
> colnames(prostate) = c(names,"TumorStatus")
```

For the purpose of demonstration, we exclude genes such that their distributions were abnormally distributed. This can be assessed by drawing a panel histograms using the function `histogram` available in the package `lattice`. We include the codes below for the purpose of plotting a panel of 100 histograms.

```
> X = prostate[,1:100]
> for (i in 1:ncol(X)){
+ if (i==1)
+ {
+ rearrangeX = cbind(X[,i],i)
+ }
+ else
+ {
```

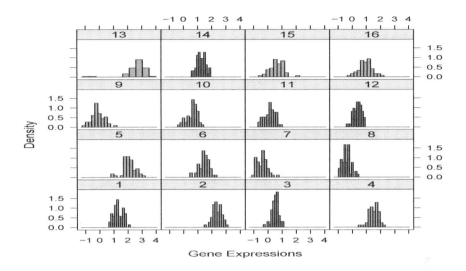

Figure 8.1   Histograms of gene expression of the 16 genes.

```
+ rearrangeX = rbind(rearrangeX,cbind(X[,i],i))
+ }}
> rearrangeX = as.data.frame(rearrangeX)
> colnames(rearrangeX) = c("GE","Gene")
> rearrangeX[,2] = as.factor(rearrangeX[,2])
> histogram(~rearrangeX[,1]|Gene,
+    data = rearrangeX)
```

Based on the distribution of the gene expression data, we selected a subset of 16 genes to infer the network. Using a program similar to the above, we draw the histogram of these 16 genes to check their potential of normality (Figure 8.1).

```
> select = c(10, 35, 40, 50, 51, 59, 61, 63, 66, 67, 68,
+ 75, 84, 89, 91, 96)
> X = prostate[,select]
```

Next, we apply the function selectFast to construct an undirected network with the $0 - 1$ conditional independence graphs (C01) and graphs constructed via the LARS-LASSO approach (LA) as candidate graphs. We extract the first 16 columns to be used by selectFast. Types of methods used to construct candidate networks are specified using family. The object generated from selectFast includes adjacency matrices from all combinations of the methods included in family; in this example, these are C01, LA and C01.LA. In the program below, we choose the adjacency matrix constructed based on graphs from both C01 and LA. In total, 12 edges are inferred. The function `graph_from_adjacency_matrix` in the `igraph` package constructs a graph based on the inferred adjacency matrix. We use `mode` to indicate the type of graphs, i.e., `undirected` in this case. By setting `diag = FALSE`, we exclude circular links. When taking `add.colnames = NULL`, the labels of each node in the graph are the generic gene names we gave earlier.

```
> # Construct a network of the genes
> library(GGMselect)
> results = selectFast(X, K = 2.5, family = c("C01","LA"))
> adjacency = results[3]$C01.LA$G
> sum(adjacency)
[1] 24
> # the package to draw a graph
> library(igraph)
> nodeNames = colnames(X)
> graphResults = graph\_from\_adjacency\_matrix(adjacency,
+ mode = "undirected", diag = FALSE, add.colnames = NULL)
```

Now we plot the graph using the function `plot.igraph`. The `igraph` package does not support postscript. To export a graph, we need to set up specific devices, e.g., `png, jpeg` or `pdf` as in the program below. We discuss several key arguments in the function `plot.igraph`. Argument `vertex.size` determines the size of each node and `vertex.label.dist` specifies the distance between the label of each node and the vertex. With `dev.off()`, we turn the device off and the figure will be saved with the specified file name, `UndirectedNetworkGE.pdf` (Figure 8.2). Note that a seed is set before plotting the graph. Without a seed, each time `plot.igraph` may generate a graph with a different format although the connections between the nodes are kept the same.

```
> windows()
> set.seed(12345)
> pdf(file = "UndirectedNetworkGE.pdf",
+   width = 6.8, height = 6.8)
> plot.igraph(graphResults,vertex.size = 5,vertex.label
+ = nodeNames,vertex.label.color = "black", frame = TRUE,
+ vertex.label.dist=1.2, vertex.color="dark gray",
+ vertex.label.cex = 0.8,edge.width = 2,edge.color = "black")
> dev.off()
windows
    1
```

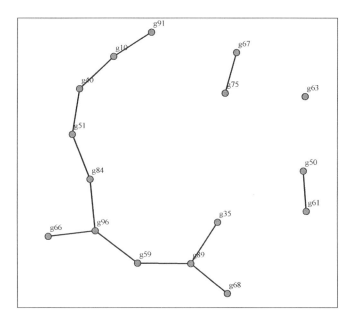

Figure 8.2 The constructed graph of genes based on gene expressions of prostate samples using `selectFast`.

Next we use `selectQE` to construct a network. Since the method in `selectQE` is based on exhaustive search, to reduce the computing time, the number of candidate graphs is limited to $5,000$, `max.nG = 5000`, which triggered the warning message after calling `selectQE`. The procedure to generate and plot the graph is the same as before. In total, 7 edges are estimated based on this setting, which results in a sparse graph (Figure 8.3). It seems graphs generated via `selectQE` are more sparse compared to graphs from `selectFast`, after testing different values of `max.nG`.

```
> results = selectQE(X, K = 2.5,max.nG = 5000)
Warning message:
In calcGrChapSymQE(SCRQE.out$Mat.Chap, SCRQE.out$SCR, scr.init, :
   *** Run stepwise procedure for iK=1 because the number of
   graphs in the collection =6355 (>max.nG=5000)
```

```
> adjacency = results$G
> sum(adjacency)
[1] 14
```

8.2 CORRELATION NETWORKS

Correlation-based networks are one type of undirected graphs and the adjacency matrices are constructed based on pair-wise correlations only. These networks have become popular in biological studies due to their simplicity and the ability to describe biological relationships among a large number of variables. In this type of network, nodes are connected based on the strength of their correlations. In this section, we focus on

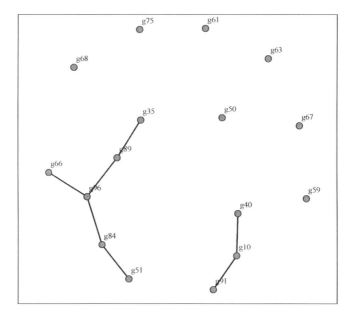

Figure 8.3 The constructed graph of genes based on gene expressions of prostate samples using selectQE.

the weighted correlation network analysis (WGCNA) proposed by Zhang and Horvath [158].

In WGCNA, the adjacency matrix is determined based on absolute values of correlation coefficients between every two nodes, denoted as co-expression similarities in Zhang and Horvath due to their focus on gene expression studies. With a pre-specified threshold τ, the matrix of co-expression similarities is converted to an adjacency matrix. That is, two nodes are linked if the absolute correlation between the two nodes is $\geq \tau$. To reflect the continuous nature of the co-expression levels in their studies, Zhang and Horvath [158] proposed approaches to construct a weighted adjacency matrix which allows the matrix entry at i-th row j-th column, a_{ij}, to take values between 0 and 1. One approach is to simply raise the similarity to a power of β,

$$a_{ij} = s_{ij}^{\beta}, \tag{8.2}$$

where s_{ij} denotes co-expression similarity at entry $\{i, j\}$ and $\beta \geq 1$ is a parameter such that the weighted adjacency a_{ij} between two nodes is proportional to their similarity at the log-scale, $\log(a_{ij}) = \beta \times \log(s_{ij})$. The parameter β needs to be set before inferring networks.

Another formulation of weighted adjacency matrix is defined by use of the sigmoid function,

$$a_{ij} = \frac{1}{1 + \exp\{-\alpha(s_{ij} - \tau_0)\}},$$

where α and τ_0 are the parameters to be specified.

In both definitions of weighted adjacency matrix, we need pre-specify parameters. The selection of those parameters is critical, as which value to choose influences the quality of the network constructed. Zhang and Horvath [158] suggest to take parameters such that approximate scale-free topology is satisfied. Scale free topology is evaluated based on connectivity for each node, k_i, defined as summation of $a_{ij}, j = 1, \cdots, n$. To visualize whether approximate scale-free topology is satisfied, one plots $\log_1 0(p(k))$ v.s $\log_{10}(k)$, where $p(k) \sim k^{-\gamma}$. If a linear relationship with negative slope is observed with, e.g., $R^2 > 0.8$, then an approximate scale-free topology can be assumed.

Densely interconnected nodes can have in depth biological interpretations. A group of such nodes is called a module. Zhang and Horvath utilize hierarchical clustering methods as well as other computing efficient approaches including block-wise module detection techniques to identify modules to be studied further on their associations with health conditions such as cancer or allergic diseases.

The methods for correlation construction and module detection, along with a number of other analytical tools have been built into a package, WGCNA[88]. In the following, we use the gene expression data discussed in Section 8.1.2 to demonstrate network construction and module selection. The data from 100 genes are used in this section, since WGCNA does not rely on normality assumptions. A detailed tutorial is available online and has examples on many other functions in the package WGCNA, https://horvath.genetics.ucla.edu/html/CoexpressionNetwork/Rpackages/WGCNA.

We first construct a correlation network using hard threshold to determine the status of connection between nodes. In this example, we choose 0.8 for the purpose of demonstration. To generate an adjacency matrix using hard threshold, function signumAdjacencyFunction is applied. Note that in the program below, X is the expression data of 100 genes extracted from the prostate data, X = prostate[,1:100]. Pearson correlations between genes are calculated using the R function cor. The program to draw the network is the same as before. Since correlation networks focus on marginal relationships, i.e., assuming independence between nodes, the overall structure (Figure 8.4) is different from what is seen based on precision matrices.

```
> library(WGCNA)
> options(stringsAsFactors = FALSE)
> # draw a hard-threshold network
> corMat = cor(X,method = "pearson")
> adj = signumAdjacencyFunction(corMat = corMat, threshold = 0.8)
```

Considering connectivity between nodes as a continuous measure, we next construct a network based on weighted adjacency matrix. A heatmap is utilized to display the adjacency between nodes. Each row and column of the heatmap is for one gene with light colors denoting low adjacency and darker colors higher adjacency.

To construct a network, the first step is to determine the parameters in the definition of weighted adjacency. The codes discussed here are adapted from the WGCNA online tutorial. The approach implemented in the WGCNA package is the weighted adjacency defined in (8.2) and parameter β needs to be specified. This is fulfilled by assessing the association of scale-free topology fit index with a set of proposed values of β, specified by powers. The scale-free topology fit index is defined as $-\text{sign}(\hat{\gamma})R^2$,

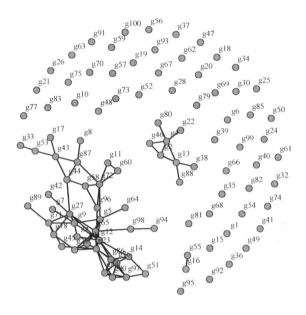

Figure 8.4 The constructed correlation network of genes based on gene expressions of prostate samples.

where γ is the regression coefficients (regressing $\log_{10}(p(k))$ on $\log_{10}(k)$) and R^2 assesses the goodness of fit. For each value in powers, function pickSoftThreshold calculates fitted regression coefficients assessing the relationship between $\log_{10}(p(k))$ and $\log_{10}(k)$, R^2, along with statistics of connectivity.

```
> powers = c(c(1:10), seq(from = 12, to = 20, by = 2))
> # call the network topology analysis function
> pickBeta = pickSoftThreshold(X, powerVector = powers, verbose = 5)
pickSoftThreshold: will use block size 100.
 pickSoftThreshold: calculating connectivity for given powers...
   ..working on genes 1 through 100 of 100
   Power SFT.R.sq  slope truncated.R.sq mean.k. median.k. max.k.
1      1   0.2620  1.560         0.5800 46.3000   47.3000 60.500
2      2   0.1970  0.746         0.8390 24.8000   25.3000 38.600
```

3	3	0.1530	0.441	0.8810	14.4000	14.0000	25.500
4	4	0.1190	0.306	0.9620	8.8600	8.1800	17.400
5	5	0.0189	-0.102	0.8340	5.6800	5.0800	12.500
6	6	0.2670	-0.371	0.8470	3.7600	3.1600	9.190
7	7	0.5690	-0.629	0.8140	2.5700	2.0500	6.860

....

We then plot scale-free topology fit indices versus the pre-specified powers (Figure 8.5). Based on the plot, the indices become stable around the power of 10, which is selected and included in the next step to infer networks and modules.

```
> cex1 = 0.9
> # Scale-free topology fit index as a function of
+ the soft-thresholding power
> plot(pickBeta$fitIndices[,1], -sign(pickBeta$fitIndices[,3])*
+ pickBeta$fitIndices[,2], xlab = "Power (beta)",
+ ylab = "Signed R^2",type = "n")
> text(pickBeta$fitIndices[,1],
+ -sign(pickBeta$fitIndices[,3])*pickBeta$fitIndices[,2],
+ labels = powers,cex = cex1,col = "black")
> abline(h = 0.80,col = "black",lty = 2)
```

The function `blockwiseModules` is applied to infer a network. The method implemented in this function is computationally efficient at the stage of detecting modules when dealing with a large number of nodes. In the following, the settings in `blockwiseModules` suggested by the WGCNA online tutorial are taken except for the values of `power` and `minModuleSize` which need to be selected based on data. Following Langfelder and Horvath [88], `TOM` is transformed with a power to make moderately strong connections more visible in the heatmap.

Sub-networks or modules in a network are helpful to the understanding of biological pathways. Furthermore, linking modules instead of each individual nodes to a health outcome or a trait reliefs the burden of multiple testing adjustment, thus helping improve statistical testing power. WGCNA utilizes clusters to identify modules. Vector `moduleLabels` has the assignment of modules to each node, which can be viewed on the heatmap as well done by `TOMplot` (Figure 8.6). Darker colors on the heatmap indicate genes with higher connectivity. To summarize each module's feature, one way is to perform principal components analyses, which is implemented in `moduleEigengenes` and the first component is used to represent the feature of a module. These inferred modules

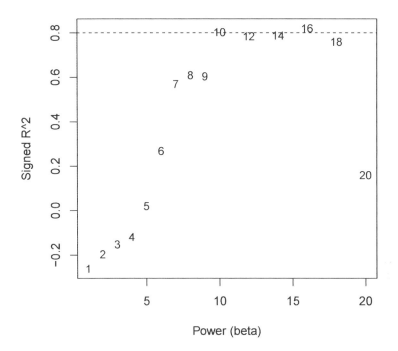

Figure 8.5 The association of scale-free topology fit index and the soft-thresholding power (β in (8.2).

can then be included in subsequent analyses for their associations with health conditions or traits of interest.

```
> netModule = blockwiseModules(X, power = 10, TOMType = "unsigned",
+ minModuleSize = 5, reassignThreshold = 0, mergeCutHeight = 0.25,
+ numericLabels = TRUE, pamRespectsDendro = FALSE, verbos = 0)
> moduleLabels = netModule$colors
> customColorOrder = c("grey", "light grey", "dark grey","black",
+ "gray");
> moduleColors = customColorOrder[moduleLabels + 1]
> network = netModule$dendrograms[[1]];
> TOM = 1-TOMsimilarityFromExpr(X, power = 10)
> plotTOM = TOM^7
> # Set diagonal to NA for a nicer plot
> diag(plotTOM) = NA
> # Call the plot function
> sizeGrWindow(9,9)
> TOMplot(plotTOM, network, moduleColors, col = gray.colors(16,
+ start = 0, end = 0.98), main = "Network heatmap plot, all genes")
```

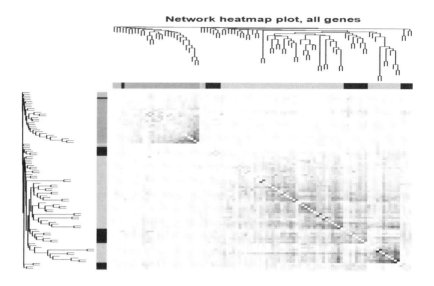

Figure 8.6 The network reflected by the weighted adjacency and the identified modules indicated by the darkness on the side bars.

8.3 BAYESIAN NETWORKS

Bayesian networks, also called probabilistic directed acyclic graphs (DAGs), are directed networks and DAGs such that edges are probabilistically connected with directions pointing from parents to children. A graph is a DAG if all the edges have directions and none of the nodes directly goes to itself or through a path to itself (a circle). Most methods to learn Bayesian networks are generally categorized into two groups, score- and constrained-based approaches, among other approaches such as the hybrids of the two approaches. In score-based approaches, a score function is defined to assess the goodness-of-fit of a network to a given data set. The goal is to identify a network to maximize the score function. This can be fulfilled through various search algorithms to search through all possible candidate graphs or through posterior sampling [21, 122, 62]. For constrained-based methods [134], a Bayesian network is inferred based on conditional independence tests between nodes. In general, this type of methods has three steps, learning Markov blankets for each node, identifying neighbors of each node, and learning direction of edge connections [125]. A Markov blanket of a node is composed of a collection of nodes needed to learn the node's parents and children.

Graphs formed based on independence tests ultimately are statistically equivalent networks consistent with the tests.

The package `bnlearn` includes a number of methods in both groups, including score- and constrained-based approaches. In this section, we focus more on the application of the package on Gaussian Bayesian networks. To demonstrate, we use the 16 genes discussed in Section 8.1 to better fit the assumptions in Gaussian Bayesian networks. We focused on two score-based approaches, `hc` and `tabu` [25, 59], and one hybrid approach `mmhc` [139]. These three functions are available in the package `bnlearn`. We start from the `hc` function and learn the structure of the network. To assess the goodness of fit with respect to the normality assumption, we draw a panel of histograms of the residuals calculated using `bn.fit`. Based on the graphs, overall we may claim that the normality assumption is met (Figure 8.7).

```
> library(bnlearn)
> Xnew = as.data.frame(X)
> dag = hc(Xnew)
> bn = bn.fit(dag,Xnew,method="mle")
> bn.fit.histogram(bn)
```

The distributions of the residuals from `tabu` and `mmhc` also roughly meet the normality assumption. However, computationally, `mmhc` is more

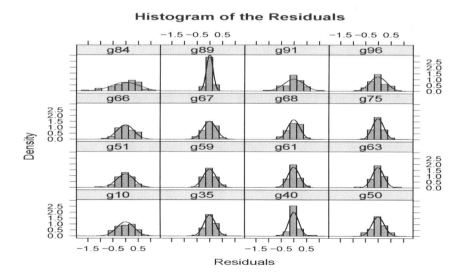

Figure 8.7 Histograms of the residuals of the 16 variables.

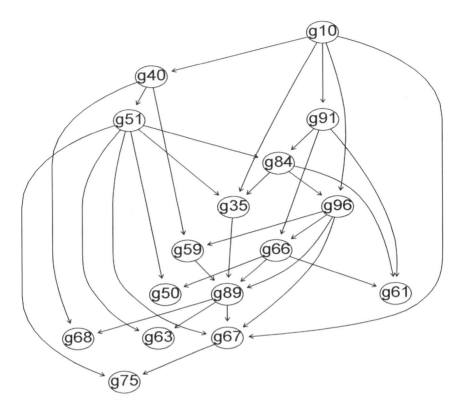

Figure 8.8 The learned graph of 16 genes using function hc.

efficient. We use function graphviz.plot in the R package Rgraphviz to draw the learned graph (Figure 8.8).

```
> library(Rgraphviz)
> graphviz.plot(dag,shape = "ellipse")
```

To reduce the uncertainty in the network learning process, we use bootstrap samples. One graph is inferred from each bootstrap sample and then we average across all the graphs. The edges will be selected based on the frequency of consistent direction of connections. Function boot.strength in the bnlearn package is implemented. This function estimates the probability of each edge being present and the probability of each edge's direction conditional on the edge being present in the graph (in either direction). In the following, we choose 500 bootstrap samples to estimate the connection strength, and in the function averaged.network used to infer the network, we take a threshold of 0.6 for the connection strength. Comparing the averaged graph

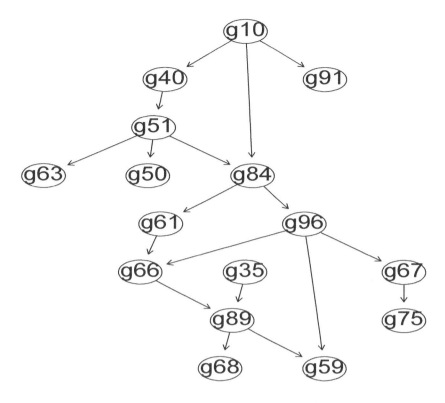

Figure 8.9 The averaged graph of 16 genes using function `hc` based on 500 bootstrap samples.

(Figure 8.9) with the initially learned graph (Figure 8.8), overall the network structure is consistent between the two graphs, and the averaged graph eliminated some edges that are highly uncertain.

```
> set.seed(12345)
> boot_hc = boot.strength(Xnew,algorithm="hc", R=500)
> avgDag = averaged.network(boot_hc,threshold=0.6)
> graphviz.plot(avgDag,shape="ellipse")
```

8.4   NETWORK COMPARISONS

In the field of network studies, most existing methods focus on inferring networks and rather limited work is on network comparisons. In this section, we discuss the progress in undirected network comparisons and Bayesian network comparisons.

8.4.1 Comparing undirected networks

For undirected graphs, several studies have proposed methods for network comparisons, most of which are conditional on known graphs. Gill et al. [56] proposed a procedure to globally test differential undirected graphs particularly applied to genes, based on strength of genetic associations or interaction between genes. Jacob et al. [78] tested multivariate two-sample means on known graphs utilizing Hotelling's T^2-tests. Some methods test graph differentiations based on comparing precision matrices, for instance, the work by Zhao et al. [164] and by Stadler et al. [135]. The method of Zhao et al. was later extended with the ability to globally test differentiation of undirected graphs [150]. Methods built upon associations of undirected networks with a feature of interest have been proposed as well [33]. A permutation-based approach has been proposed [141], which has the ability to compare network structure, global strength, and edge connection strength. For Gaussian undirected networks, the networks are constructed based on l_1-regularized partial correlations. Among all these methods, the approach by van Borkulo et al. [141] has been implemented in the package `NetworkComparisonTest` and is discussed in this section. Since this approach is built upon permutations, it is understandable that this approach has computing challenge when dealing with large numbers of variables.

Recall that in the gene expression data of prostate samples discussed so far in this chapter, 25 samples are tumorous and the remaining are non-tumorous. Let us apply `NetworkComparisonTest` to examine the differentiation in joint activities among genes between the two groups. We use the gene expression data of 16 genes to compare the networks with respect to their global difference. The R codes are given below, in which the function `NCT` is used for the test. Most arguments in `NCT` are self-explained and we give a brief explanation on `test.edges` and `edges`. Argument `test.edges` in `NCT` is to indicate whether differences between individual edges should be tested. Setting `edges = "all"` is to indicate the comparison is based on all the edges. If a subset of edges is of particular interest in the comparison, then a list of paired indices referring to the corresponding nodes should be provided for `edges`. Based on this permutation test on gene expressions, the networks of these 16 genes are not different between the two groups (p-value is 1).

```
> library(NetworkComparisonTest)
> data1 = X[1:25,]
> data2 = X[26:50,]
```

```
> NCT(data1, data2, it = 100, test.edges = FALSE, edges = "all",
+ progressbar = FALSE, p.adjust.methods = "fdr")
 NETWORK INVARIANCE TEST
 Test statistic M:  0
 p-value 1
 GLOBAL STRENGTH INVARIANCE TEST
 Global strength per group:  0 0
 Test statistic S:  0
 p-value 1
 EDGE INVARIANCE TEST
NULL
 CENTRALITY INVARIANCE TEST
NULL
```

8.4.2   Comparing Bayesian networks

Approaches that can both construct Bayesian networks and statistically test for differentiation between Bayesian networks are lacking, but the bnlearn package has functions that are able to do some simple comparisons with respect to the agreement in edges between the networks assuming the inferred networks are true networks. We continue to use data1 and data2 discussed above in the comparison between undirected graphs.

We use hc to learn the Bayesian networks for tumorous samples and non-tumorous samples, respectively. The function to compare the two networks is all.equal. As seen in the output from all.equal, this function treats both inferred networks as the truth and examines if the number of edges agrees with each other. It concludes that the two networks are different based on a simple comparison in the connections between the networks.

```
> dag1 = hc(as.data.frame(data1))
> dag2 = hc(as.data.frame(data2))
> all.equal(dag1,dag2)
[1] "Different number of directed/undirected arcs"
```

Another comparison is to treat one of the inferred Bayesian network as the truth or the "target" network, and compare the other network with the target. The function to fulfill this task is compare. The first network object is treated as the target, and the second network is compared to the target to identify the number of edges that are true positives (tp), false positives (fp), and false negatives (fn).

```
> compare(dag1,dag2)
```

```
$tp
[1] 5
$fp
[1] 24
$fn
[1] 23
```

As seen from the above examples, these conclusions are based on simple counting with `dag1` as the target. It is desirable to have more rigorous comparisons, e.g., a statistical testing to test whether two Bayesian networks are the same, and identify potential driving variables leading to the differentiation. More studies and investigations are certainly needed in this area.

Bibliography

[1] Affymetrix. statistical algorithms description document. Technical report.

[2] Michael G Akritas and Nikolaos Papadatos. Heteroscedastic one-way anova and lack-of-fit tests. *Journal of the American Statistical Association*, 99(466):368–382, 2004.

[3] Charles E Antoniak. Mixtures of dirichlet processes with applications to bayesian nonparametric problems. *The annals of statistics*, pages 1152–1174, 1974.

[4] Kelly M Bakulski, Jason I Feinberg, Shan V Andrews, Jack Yang, Shannon Brown, Stephanie L. McKenney, Frank Witter, Jeremy Walston, Andrew P Feinberg, and M Daniele Fallin. Dna methylation of cord blood cell types: applications for mixed cell birth studies. *Epigenetics*, 11(5):354–362, 2016.

[5] M Bartlett. A comment on D. V. Lindley's statistical paradox. *Biometrika*, 44:533–534, 1957.

[6] Maria J Bayarri, James O Berger, Anabel Forte, G García-Donato, et al. Criteria for bayesian model choice with application to variable selection. *The Annals of statistics*, 40(3):1550–1577, 2012.

[7] Yoav Benjamini and Yosef Hochberg. Controlling the false discovery rate: a practical and powerful approach to multiple testing. *Journal of the Royal statistical society: series B (Methodological)*, 57(1):289–300, 1995.

[8] Yoav Benjamini, Daniel Yekutieli, et al. The control of the false discovery rate in multiple testing under dependency. *The annals of statistics*, 29(4):1165–1188, 2001.

[9] Ben Bolstad. affyplm: the threestep function, 2016.

[10] Benjamin M Bolstad, Rafael A Irizarry, Magnus Åstrand, and Terence P. Speed. A comparison of normalization methods for high density oligonucleotide array data based on variance and bias. *Bioinformatics*, 19(2):185–193, 2003.

[11] Leo Breiman. Bagging predictors. *Machine learning*, 24(2):123–140, 1996.

[12] Leo Breiman. Random forests. *Machine learning*, 45(1):5–32, 2001.

[13] Angela N Brooks, Li Yang, Michael O Duff, Kasper D Hansen, Jung W Park, Sandrine Dudoit, Steven E Brenner, and Brenton R Graveley. Conservation of an rna regulatory map between drosophila and mammals. *Genome research*, 21(2):193–202, 2011.

[14] Patrick O Brown and David Botstein. Exploring the new world of the genome with dna microarrays. *Nature genetics*, 21(1s):33, 1999.

[15] Andreas Buja and Nermin Eyuboglu. Remarks on parallel analysis. *Multivariate behavioral research*, 27(4):509–540, 1992.

[16] Emmanuel Candes, Terence Tao, et al. The dantzig selector: Statistical estimation when p is much larger than n. *The annals of Statistics*, 35(6):2313–2351, 2007.

[17] Robert Castelo and Alberto Roverato. A robust procedure for gaussian graphical model search from microarray data with p larger than n. *Journal of Machine Learning Research*, 7(Dec):2621–2650, 2006.

[18] J L Castle and D F Hendry. Automatic selection for non-linear models. In H Garnier L Wang and T Jackman, editors, *System Identification, Environmental Modelling and Control*. Springer Verlag, 2010.

[19] Jiahua Chen and Zehua Chen. Extended bayesian information criteria for model selection with large model spaces. *Biometrika*, 95(3):759–771, 2008.

[20] Hugh Chipman and Robert Tibshirani. Hybrid hierarchical clustering with applications to microarray data. *Biostatistics*, 7(2):286–301, 2006.

[21] Gregory F Cooper and Edward Herskovits. A bayesian method for the induction of probabilistic networks from data. *Machine learning*, 9(4):309–347, 1992.

[22] Douglas R Cox. Note on grouping. *Journal of the American Statistical Association*, 52(280):543–547, 1957.

[23] N Cristianini and J Shawe-Taylor. *An Introduction to Support Vector Machines and Other Kernel-based Learning Methods*. Cambridge University Press, 2000.

[24] A Dalalyan and A Tsybakov. Aggregation by exponential weighting, sharp oracle inequalities and sparsity. *Machine Learning*, 72(1-2):39–61, 2008.

[25] Rónán Daly and Qiang Shen. Methods to accelerate the learning of bayesian network structures. In *Proceedings of the 2007 UK Workshop on Computational Intelligence*, 2007.

[26] Ramón Díaz-Uriarte and Sara Alvarez De Andres. Gene selection and classification of microarray data using random forest. *BMC bioinformatics*, 7(1):3, 2006.

[27] Kevin K Dobbin and Richard M Simon. Optimally splitting cases for training and testing high dimensional classifiers. *BMC medical genomics*, 4(1):31, 2011.

[28] Pan Du, Xiao Zhang, Chiang-Ching Huang, Nadereh Jafari, Warren A Kibbe, Lifang Hou, and Simon M Lin. Comparison of beta-value and m-value methods for quantifying methylation levels by microarray analysis. *BMC bioinformatics*, 11(1):587, 2010.

[29] Sandrine Dudoit and Jane Fridlyand. Classification in microarray experiments. *Statistical analysis of gene expression microarray data*, 1:93–158, 2003.

[30] Sandrine Dudoit, Jane Fridlyand, and Terence P Speed. Comparison of discrimination methods for the classification of tumors using gene expression data. *Journal of the American statistical association*, 97(457):77–87, 2002.

[31] Olive Jean Dunn. Estimation of the medians for dependent variables. *The Annals of Mathematical Statistics*, pages 192–197, 1959.

[32] Olive Jean Dunn. Multiple comparisons among means. *Journal of the American statistical association*, 56(293):52–64, 1961.

[33] Daniele Durante, David B Dunson, et al. Bayesian inference and testing of group differences in brain networks. *Bayesian analysis*, 13(1):29–58, 2018.

[34] David Edwards. Non-linear normalization and background correction in one-channel cdna microarray studies. *Bioinformatics*, 19(7):825–833, 2003.

[35] Bradley Efron, Trevor Hastie, Iain Johnstone, Robert Tibshirani, et al. Least angle regression. *The Annals of statistics*, 32(2):407–499, 2004.

[36] Paul H C Eilers and Brian D Marx. Flexible smoothing with B-splines and penalties. *Statistical Science*, 11:89–102, 1996.

[37] Jianqing Fan and Runze Li. Variable selection via nonconcave penalized likelihood and its oracle properties. *Journal of the American statistical Association*, 96(456):1348–1360, 2001.

[38] Jianqing Fan and Jinchi Lv. Sure independence screening for ultra-high dimensional feature space. *Journal of the Royal Statistical Society: Series B (Statistical Methodology)*, 70(5):849–911, 2008.

[39] C Fernández, E Ley, and M F J Steel. Benchmark priors for bayesian model averaging. *Journal of Econometrics*, 100(2):381–427, 2001.

[40] Stephen P Fodor, J Leighton Read, Michael C Pirrung, Lubert Stryer, A Tsai Lu, and Dennis Solas. Light-directed, spatially addressable parallel chemical synthesis. *science*, 251(4995):767–773, 1991.

[41] Edward Forgey. Cluster analysis of multivariate data: Efficiency vs. interpretability of classification. *Biometrics*, 21(3):768–769, 1965.

[42] Jean-Philippe Fortin, Timothy J Triche, and Kasper D Hansen. Preprocessing, normalization and integration of the illumina humanmethylationepic array with minfi. *Bioinformatics*, 33:558–560, 2017.

[43] D P Foster and E I George. The risk inflation criterion for multiple regression. *Annals of Statistics*, 22:1947–1975, 1994.

[44] Luis Ángel García-Escudero and Alfonso Gordaliza. Robustness properties of k means and trimmed k means. *Journal of the American Statistical Association*, 94(447):956–969, 1999.

[45] Christopher Genovese, Jiashun Jin, and Larry Wasserman. Revisiting marginal regression. *arXiv preprint arXiv:0911.4080*, 2009.

[46] Robert Gentleman, Vincent Carey, Wolfgang Huber, Rafael Irizarry, and Sandrine Dudoit. *Bioinformatics and computational biology solutions using R and Bioconductor*. Springer Science & Business Media, 2006.

[47] Robin Genuer, Jean-Michel Poggi, and Christine Tuleau-Malot. Variable selection using random forests. *Pattern Recognition Letters*, 31(14):2225–2236, 2010.

[48] E I George. The variable selection problem. *Journal of the American Statistical Association*, 95:1304–1308, 2000.

[49] E I George and D P Foster. Calibration and empirical Bayes variable selection. *Biometrika*, 87:731–747, 2000.

[50] E I George and R E McCulloch. Approaches for Bayesian variable selection. *Statistica Sinica*, 7:339–374, 1997.

[51] E I George and R E McCulloch. Comment on "Variable selection and function estimation in additive nonparametric regression using a data-based prior". *Journal of the American Statistical Association*, 94:798–799, 1999.

[52] Edward I George and Robert E McCulloch. Variable selection via gibbs sampling. *Journal of the American Statistical Association*, 88(423):881–889, 1993.

[53] Alexandros Georgogiannis. Robust k-means: a theoretical revisit. In *Advances in Neural Information Processing Systems*, pages 2891–2899, 2016.

[54] Allen Gersho and Robert M Gray. *Vector quantization and signal compression*, volume 159. Springer Science & Business Media, 2012.

[55] D Gianola and J B van Kaam. Reproducing kernel Hilbert spaces regression methods for genomic assisted prediction of quantitative traits. *Genetics*, 178:2289–2303, 2008.

[56] Ryan Gill, Somnath Datta, and Susmita Datta. A statistical framework for differential network analysis from microarray data. *BMC bioinformatics*, 11(1):1, 2010.

[57] Christophe Giraud. Estimation of gaussian graphs by model selection. *Electronic Journal of Statistics*, 2:542–563, 2008.

[58] Christophe Giraud, Sylvie Huet, and Nicolas Verzelen. Graph selection with ggmselect. *Statistical applications in genetics and molecular biology*, 11(3), 2012.

[59] Fred Glover. Future paths for integer programming and links to artificial intelligence. *Computers & operations research*, 13(5):533–549, 1986.

[60] John C Gower. A general coefficient of similarity and some of its properties. *Biometrics*, pages 857–871, 1971.

[61] Pablo M Granitto, Cesare Furlanello, Franco Biasioli, and Flavia Gasperi. Recursive feature elimination with random forest for ptr-ms analysis of agroindustrial products. *Chemometrics and Intelligent Laboratory Systems*, 83(2):83 – 90, 2006.

[62] Shengtong Han, Hongmei Zhang, Ramin Homayouni, and Wilfried Karmaus. An efficient Bayesian approach for Gaussian bayesian network structure learning. *Communications in Statistics-Simulation and Computation*, 2016.

[63] Shengtong Han, Hongmei Zhang, Wenhui Sheng, and Hasan Arshad. The nested joint clustering via dirichlet process mixture model. *Journal of Statistical Computation and Simulation*, 89(5):815–830, 2019.

[64] John A Hartigan and Manchek A Wong. Algorithm as 136: A k-means clustering algorithm. *Journal of the Royal Statistical Society. Series C (Applied Statistics)*, 28(1):100–108, 1979.

[65] Ville Hautamäki, Svetlana Cherednichenko, Ismo Kärkkäinen, Tomi Kinnunen, and Pasi Fränti. Improving k-means by outlier

removal. In *Scandinavian Conference on Image Analysis*, pages 978–987. Springer, 2005.

[66] Yosef Hochberg and Yoav Benjamini. More powerful procedures for multiple significance testing. *Statistics in medicine*, 9(7):811–818, 1990.

[67] Arthur E Hoerl and Robert W Kennard. Ridge regression: Biased estimation for nonorthogonal problems. *Technometrics*, 12(1):55–67, 1970.

[68] Katsuhiro Honda, Akira Notsu, and Hidetomo Ichihashi. Fuzzy pca-guided robust k-means clustering. *IEEE Transactions on Fuzzy Systems*, 18(1):67–79, 2010.

[69] Torsten Hothorn and Brian S Everitt. *A handbook of statistical analyses using R*. Chapman and Hall/CRC, 2009.

[70] E Andres Houseman, Molly L Kile, David C Christiani, Tan A Ince, Karl T Kelsey, and Carmen J Marsit. Reference-free deconvolution of dna methylation data and mediation by cell composition effects. *BMC bioinformatics*, 17(1):259, 2016.

[71] Eugene Andres Houseman, William P Accomando, Devin C Koestler, Brock C Christensen, Carmen J Marsit, Heather H Nelson, John K Wiencke, and Karl T Kelsey. Dna methylation arrays as surrogate measures of cell mixture distribution. *BMC bioinformatics*, 13(1):86, 2012.

[72] Eugene Andres Houseman, John Molitor, and Carmen J Marsit. Reference-free cell mixture adjustments in analysis of dna methylation data. *Bioinformatics*, 30(10):1431–1439, 2014.

[73] Rafael A Irizarry, Bridget Hobbs, Francois Collin, Yasmin D Beazer-Barclay, Kristen J Antonellis, Uwe Scherf, and Terence P Speed. Exploration, normalization, and summaries of high density oligonucleotide array probe level data. *Biostatistics*, 4(2):249–264, 2003.

[74] H Ishwaran and J S Rao. Spike and slab gene selection for multigroup microarray data. *Journal of the American Statistical Association*, 100:764–780, 2005.

[75] H Ishwaran and J S Rao. Spike and slab variable selection: Frequentist and Bayesian strategies. *The Annals of Statistics*, 33:730–773, 2005.

[76] H Ishwaran and J S Rao. Generalized ridge regression: geometry and computational solutions when p is larger than n, 2010.

[77] Hemant Ishwaran, Udaya B Kogalur, and J Sunil Rao. spikeslab: Prediction and variable selection using spike and slab regression. *R Journal*, 2(2), 2010.

[78] Laurent Jacob, Pierre Neuvial, and Sandrine Dudoit. More power via graph-structured tests for differential expression of gene networks. *The Annals of Applied Statistics*, pages 561–600, 2012.

[79] Andrew E Jaffe and Rafael A Irizarry. Accounting for cellular heterogeneity is critical in epigenome-wide association studies. *Genome biology*, 15(2):R31, 2014.

[80] Yu Jiang, Jinfeng Wei, Hongmei Zhang, Susan Ewart, Faisal I Rezwan, John W Holloway, Hasan Arshad, and Wilfried Karmaus. Epigenome wide comparison of dna methylation profile between paired umbilical cord blood and neonatal blood on guthrie cards. *Epigenetics*, pages 1–8, 2019.

[81] David Richard Johnson and James C Creech. Ordinal measures in multiple indicator models: A simulation study of categorization error. *American Sociological Review*, pages 398–407, 1983.

[82] Bonnie R Joubert, Siri E Håberg, Roy M Nilsen, Xuting Wang, Stein E Vollset, Susan K Murphy, Zhiqing Huang, Cathrine Hoyo, Øivind Midttun, Lea A Cupul-Uicab, et al. 450k epigenome-wide scan identifies differential dna methylation in newborns related to maternal smoking during pregnancy. *Environmental health perspectives*, 120(10):1425–1431, 2012.

[83] Leonard Kaufman and Peter Rousseeuw. *Clustering by means of medoids*. North-Holland, 1987.

[84] Leonard Kaufman and Peter J Rousseeuw. *Finding groups in data: an introduction to cluster analysis*, volume 344. John Wiley & Sons, 2009.

[85] Akhilesh Kaushal, Hongmei Zhang, Wilfried J J Karmaus, Meredith Ray, Mylin A Torres, Alicia K Smith, and Shu-Li Wang. Comparison of different cell type correction methods for genome-scale epigenetics studies. *Bmc Bioinformatics*, 18(1):216, 2017.

[86] Vladimir Kogan, Joshua Millstein, Stephanie J London, Carole Ober, Steven R White, Edward T Naureckas, W James Gauderman, Daniel J Jackson, Albino Barraza-Villarreal, Isabelle Romieu, et al. Genetic-epigenetic interactions in asthma revealed by a genome-wide gene-centric search. *Human heredity*, 83(3):130–152, 2018.

[87] Yutaka Kondo. Epigenetic cross-talk between dna methylation and histone modifications in human cancers. *Yonsei medical journal*, 50(4):455–463, 2009.

[88] Peter Langfelder and Steve Horvath. WGCNA: an r package for weighted correlation network analysis. *BMC bioinformatics*, 9(1):559, 2008.

[89] Laura Lazzeroni and Art Owen. Plaid models for gene expression data. *Statistica sinica*, pages 61–86, 2002.

[90] K E Lee, N Sha, E R Dougherty, M Vannucci, and B K Mallick. Gene selection: a Bayesian variable selection approach. *Bioinformatics*, 19:90–97, 2003.

[91] Jeffrey T Leek. Asymptotic conditional singular value decomposition for high-dimensional genomic data. *Biometrics*, 67(2):344–352, 2011.

[92] Jeffrey T Leek, W Evan Johnson, Hilary S Parker, Andrew E Jaffe, and John D Storey. *sva: Surrogate Variable Analysis*, 2014. R package version 3.8.0.

[93] Jeffrey T Leek and John D Storey. Capturing heterogeneity in gene expression studies by surrogate variable analysis. *PLoS genetics*, 3(9):e161, 2007.

[94] J T Leek, W E Johnson, H S Parker, E J Fertig, A E Jaffe, J D Storey, Y Zhang, and L C Torres. sva: Surrogate variable analysis. *R package version 3.32.1*, 2019.

[95] Benjamin Lehne, Alexander W Drong, Marie Loh, Weihua Zhang, William R Scott, Sian-Tsung Tan, Uzma Afzal, James Scott, Marjo-Riitta Jarvelin, Paul Elliott, et al. A coherent approach for analysis of the illumina humanmethylation450 beadchip improves data quality and performance in epigenome-wide association studies. *Genome biology*, 16(1):37, 2015.

[96] Jun Li and Robert Tibshirani. Finding consistent patterns: a nonparametric approach for identifying differential expression in RNA-Seq data. *Statistical methods in medical research*, 22(5):519–536, 2013.

[97] F Liang, R Paulo, G Molina, M A Clyde, and J O Berger. Mixtures of g priors for Bayesian variable selection. *Journal of the American Statistical Association*, 103:410–423, 2008.

[98] Andy Liaw, Matthew Wiener, et al. Classification and regression by randomforest. *R news*, 2(3):18–22, 2002.

[99] Yi Lin, Hao Helen Zhang, et al. Component selection and smoothing in multivariate nonparametric regression. *The Annals of Statistics*, 34(5):2272–2297, 2006.

[100] D V Lindley. A statistical paradox. *Biometrika*, 44:187–192, 1957.

[101] Miao Liu, Mingjun Wang, Jun Wang, and Duo Li. Comparison of random forest, support vector machine and back propagation neural network for electronic tongue data classification: Application to the recognition of orange beverage and chinese vinegar. *Sensors and Actuators B: Chemical*, 177:970–980, 2013.

[102] S P Lloyd. Least square quantization in pcm. bell telephone laboratories paper. published in journal much later: Lloyd, sp: Least squares quantization in pcm. *IEEE Trans. Inform. Theor.(1957/1982)*, 1957.

[103] David J Lockhart, Helin Dong, Michael C Byrne, Maximillian T Follettie, Michael V Gallo, Mark S Chee, Michael Mittmann, Chunwei Wang, Michiko Kobayashi, Heidi Norton, et al. Expression monitoring by hybridization to high-density oligonucleotide arrays. *Nature biotechnology*, 14(13):1675, 1996.

[104] Zichen Ma and Ernest Fokoue. *Preprint on arXiv, arXiv:1503.06370v2*, 03 2016.

[105] James MacQueen et al. Some methods for classification and analysis of multivariate observations. In *Proceedings of the fifth Berkeley symposium on mathematical statistics and probability*, volume 1, pages 281–297. Oakland, CA, USA, 1967.

[106] Jovana Maksimovic, Johann A Gagnon-Bartsch, Terence P Speed, and Alicia Oshlack. Removing unwanted variation in a differential methylation analysis of illumina humanmethylation450 array data. *Nucleic acids research*, 43(16):e106–e106, 2015.

[107] Monnie McGee and Zhongxue Chen. Parameter estimation for the exponential-normal convolution model for background correction of affymetrix genechip data. *Statistical applications in genetics and molecular biology*, 5(1), 2006.

[108] Nicolai Meinshausen, Peter Bühlmann, et al. High-dimensional graphs and variable selection with the lasso. *The annals of statistics*, 34(3):1436–1462, 2006.

[109] J Mercer. Functions of positive and negative type, and their connection with the theory of integral equations. *Philosophical Transactions of the Royal Society of London. Series A*, 209:415–446, 1909.

[110] T J Mitchell and J J Beauchamp. Bayesian variable selection in linear regression (C/R: P1033-1036). *Journal of the American Statistical Association*, 83:1023–1032, 1988.

[111] Frederick Mosteller and John Wilder Tukey. Data analysis and regression: a second course in statistics. *Addison-Wesley Series in Behavioral Science: Quantitative Methods*, 1977.

[112] Tucker A Patterson, Edward K Lobenhofer, Stephanie B Fulmer-Smentek, Patrick J Collins, Tzu-Ming Chu, Wenjun Bao, Hong Fang, Ernest S Kawasaki, Janet Hager, Irina R Tikhonova, et al. Performance comparison of one-color and two-color platforms within the microarray quality control (maqc) project. *Nature biotechnology*, 24(9):1140, 2006.

[113] Amela Prelić, Stefan Bleuler, Philip Zimmermann, Anja Wille, Peter Bühlmann, Wilhelm Gruissem, Lars Hennig, Lothar Thiele, and Eckart Zitzler. A systematic comparison and evaluation of

biclustering methods for gene expression data. *Bioinformatics*, 22(9):1122–1129, 2006.

[114] Li-Xuan Qin, Linda Breeden, and Steven G Self. Finding gene clusters for a replicated time course study. *BMC research notes*, 7(1):60, 2014.

[115] Li-Xuan Qin and Steven G Self. The clustering of regression models method with applications in gene expression data. *Biometrics*, 62(2):526–533, 2006.

[116] P Radchenko and G M James. Variable selection using adaptive nonlinear interaction structures in high dimensions. *Journal of the American Statistical Association*, 104:1541–1553, 2010.

[117] Elior Rahmani, Noah Zaitlen, Yael Baran, Celeste Eng, Donglei Hu, Joshua Galanter, Sam Oh, Esteban G Burchard, Eleazar Eskin, James Zou, et al. Sparse pca corrects for cell type heterogeneity in epigenome-wide association studies. *Nature methods*, 13(5):443, 2016.

[118] Meredith A Ray, Xin Tong, Gabrielle A Lockett, Hongmei Zhang, and Wilfried JJ Karmaus. An efficient approach to screening epigenome-wide data. *BioMed research international*, 2016, 2016.

[119] B D Ripley. Modern applied statistics with s. *Statistics and Computing, fourth ed. Springer, New York*, 2002.

[120] Matthew E Ritchie, Jeremy Silver, Alicia Oshlack, Melissa Holmes, Dileepa Diyagama, Andrew Holloway, and Gordon K Smyth. A comparison of background correction methods for two-colour microarrays. *Bioinformatics*, 23(20):2700–2707, 2007.

[121] L Rosasco, S Mosci, M S Santoro, A Verri, and S Villa. A regularization approach to nonlinear variable selection. In *Proceedings of the 13 International Conference on Artificial Intelligence and Statistics*, 2010.

[122] Stuart J Russell and Peter Norvig. *Artificial intelligence: a modern approach*. Malaysia; Pearson Education Limited,, 2016.

[123] Diego Franco Saldana and Yang Feng. Sis: An r package for sure independence screening in ultrahigh dimensional statistical models. *Journal of Statistical Software*, 83, 2018.

[124] James G Scott and Carlos M Carvalho. Feature-inclusion stochastic search for gaussian graphical models. *Journal of Computational and Graphical Statistics*, 17(4):790–808, 2008.

[125] Marco Scutari. Bayesian network constraint-based structure learning algorithms: Parallel and optimized implementations in the bnlearn R package. *Journal of Statistical Software*, 077(i02), 2017.

[126] Marta Sestelo, Nora M Villanueva, Luis Meira-Machado, and Javier Roca-Pardiñas. Fwdselect: An r package for variable selection in regression models. *R Journal*, 8(1), 2016.

[127] Jiejun Shi and Li-Xuan Qin. Corm: An r package implementing the clustering of regression models method for gene clustering: Supplement issue: Array platform modeling and analysis (a). *Cancer informatics*, 13:CIN–S13967, 2014.

[128] Jeremy D Silver, Matthew E Ritchie, and Gordon K Smyth. Microarray background correction: maximum likelihood estimation for the normal–exponential convolution. *Biostatistics*, page kxn042, 2009.

[129] Dinesh Singh, Phillip G Febbo, Kenneth Ross, Donald G Jackson, Judith Manola, Christine Ladd, Pablo Tamayo, Andrew A Renshaw, Anthony V D'Amico, Jerome P Richie, et al. Gene expression correlates of clinical prostate cancer behavior. *Cancer cell*, 1(2):203–209, 2002.

[130] M l Smith and R. Kohn. Nonparametric regression using Bayesian variable selection. *Journal of Econometrics*, 75:317–343, 1996.

[131] Smyth Gordon K. Linear models and empirical Bayes methods for assessing differential expression in microarray experiments. *Statistical Applications in Genetics and Molecular Biology*, 3(1):1–25, 2004.

[132] Agniva Som, Christopher M Hans, and Steven N MacEachern. Block hyper-g priors in bayesian regression. *arXiv preprint arXiv:1406.6419*, 2014.

[133] Nelís Soto-Ramírez, Syed Hasan Arshad, John W Holloway, Hongmei Zhang, Eric Schauberger, Susan Ewart, Veeresh Patil, and Wilfried Karmaus. The interaction of genetic variants and dna

methylation of the interleukin-4 receptor gene increase the risk of asthma at age 18 years. *Clinical epigenetics*, 5(1):1, 2013.

[134] Peter Spirtes, Clark N Glymour, Richard Scheines, David Heckerman, Christopher Meek, Gregory Cooper, and Thomas Richardson. *Causation, prediction, and search*. MIT press, 2000.

[135] Nicolas Städler, Frank Dondelinger, Steven M Hill, Rehan Akbani, Yiling Lu, Gordon B Mills, and Sach Mukherjee. Molecular heterogeneity at the network level: high-dimensional testing, clustering and a tcga case study. *Bioinformatics*, 33(18):2890–2896, 2017.

[136] Alexander Statnikov, Lily Wang, and Constantin F. Aliferis. A comprehensive comparison of random forests and support vector machines for microarray-based cancer classification. *BMC Bioinformatics*, 9(1):319, Jul 2008.

[137] William Terry, Hongmei Zhang, Arnab Maity, Hasan Arshad, and Wilfried Karmaus. Unified variable selection in semi-parametric models. *Statistical methods in medical research*, 26(6):2821–2831, 2017.

[138] Robert Tibshirani. Regression shrinkage and selection via the lasso. *Journal of the Royal Statistical Society: Series B (Methodological)*, 58(1):267–288, 1996.

[139] Ioannis Tsamardinos, Laura E Brown, and Constantin F Aliferis. The max-min hill-climbing bayesian network structure learning algorithm. *Machine learning*, 65(1):31–78, 2006.

[140] Heather Turner, Trevor Bailey, and Wojtek Krzanowski. Improved biclustering of microarray data demonstrated through systematic performance tests. *Computational statistics & data analysis*, 48(2):235–254, 2005.

[141] Claudia D van Borkulo, Lynn Boschloo, J Kossakowski, Pia Tio, Robert A Schoevers, Denny Borsboom, and Lourens J Waldorp. Comparing network structures on three aspects: A permutation test. *Manuscript submitted for publication*, 2017.

[142] Mark J Van der Laan and Katherine S Pollard. A new algorithm for hybrid hierarchical clustering with visualization and the bootstrap. *Journal of Statistical Planning and Inference*, 117(2):275–303, 2003.

[143] Vladimir Vapnik. *Statistical learning theory. 1998*, volume 3. Wiley, New York, 1998.

[144] Ting Wang, Zhao Ren, Ying Ding, Zhou Fang, Zhe Sun, Matthew L MacDonald, Robert A Sweet, Jieru Wang, and Wei Chen. Fastggm: an efficient algorithm for the inference of gaussian graphical model in biological networks. *PLoS computational biology*, 12(2):e1004755, 2016.

[145] Zhong Wang, Mark Gerstein, and Michael Snyder. RNA-Seq: a revolutionary tool for transcriptomics. *Nature reviews genetics*, 10(1):57–63, 2009.

[146] Joe Whittaker. *Graphical models in applied multivariate statistics*. Wiley Publishing, 2009.

[147] Anja Wille and Peter Bühlmann. Low-order conditional independence graphs for inferring genetic networks. *Statistical applications in genetics and molecular biology*, 5(1), 2006.

[148] Ann S Wilson, Bridget G Hobbs, Terence P Speed, and P Elizabeth Rakoczy. The microarray: potential applications for ophthalmic research. *Mol Vis*, 8:259–70, 2002.

[149] Christopher Workman, Lars Juhl Jensen, Hanne Jarmer, Randy Berka, Laurent Gautier, Henrik Bjørn Nielser, Hans-Henrik Saxild, Claus Nielsen, Søren Brunak, and Steen Knudsen. A new non-linear normalization method for reducing variability in dna microarray experiments. *Genome biol*, 3(9):1–16, 2002.

[150] Yin Xia, Tianxi Cai, and T Tony Cai. Testing differential networks with applications to the detection of gene-gene interactions. *Biometrika*, page asu074, 2015.

[151] Loïc Yengo, Julien Jacques, Christophe Biernacki, and Mickael Canouil. Variable clustering in high-dimensional linear regression: The r package clere. *The R Journal*, 8(1):92–106, 2016.

[152] Ming Yuan and Yi Lin. Model selection and estimation in the gaussian graphical model. *Biometrika*, 94(1):19–35, 2007.

[153] Adriano Zanin Zambom and Michael G Akritas. Nonparametric lack-of-fit testing and consistent variable selection. *Statistica Sinica*, pages 1837–1858, 2014.

[154] Adriano Zanin Zambom, Michael G Akritas, et al. Nonpmodelcheck: An r package for nonparametric lack-of-fit testing and variable selection. *Journal of Statistical Software*, 77(10):1–28, 2017.

[155] A Zellner. Basic issues in econometrics, chicago: U, 1984.

[156] A Zellner. On assessing prior distributions and Bayesian regression analysis with g-prior distributions. In Prem K. Goel and Arnold Zellner, editors, *Bayesian Inference and Decision Techniques: Essays in Honor of Bruno de Finetti*, pages 233–243. Elsevier/North-Holland [Elsevier Science Publishing Co., New York; North-Holland Publishing Co., Amsterdam], 1986.

[157] Arnold Zellner and Aloysius Siow. Posterior odds ratios for selected regression hypotheses. *Trabajos de estadística y de investigación operativa*, 31(1):585–603, 1980.

[158] Bin Zhang and Steve Horvath. A general framework for weighted gene co-expression network analysis. *Statistical applications in genetics and molecular biology*, 4(1), 2005.

[159] Hongmei Zhang, Xianzheng Huang, Jianjun Gan, Wilfried Karmaus, Tara Sabo-Attwood, et al. A two-component g-prior for variable selection. *Bayesian Analysis*, 11(2):353–380, 2016.

[160] Hongmei Zhang, Arnab Maity, Hasan Arshad, John Holloway, and Wilfried Karmaus. Variable selection in semi-parametric models. *Statistical methods in medical research*, 25(4):1736–1752, 2016.

[161] Hongmei Zhang, Xin Tong, John W Holloway, Faisal I Rezwan, Gabrielle A Lockett, Veeresh Patil, Meredith Ray, Todd M Everson, Nelís Soto-Ramírez, S Hasan Arshad, et al. The interplay of dna methylation over time with th2 pathway genetic variants on asthma risk and temporal asthma transition. *Clinical epigenetics*, 6(1):1, 2014.

[162] Hongmei Zhang, Yubo Zou, Will Terry, Wilfried Karmaus, and Hasan Arshad. Joint clustering with correlated variables. *The American Statistician*, pages 1–11, 2018.

[163] Lue Ping Zhao, Ross Prentice, and Linda Breeden. Statistical modeling of large microarray data sets to identify stimulus-

response profiles. *Proceedings of the National Academy of Sciences*, 98(10):5631–5636, 2001.

[164] Sihai Dave Zhao, T Tony Cai, and Hongzhe Li. Direct estimation of differential networks. *Biometrika*, 101(2):253–268, 2014.

[165] Hui Zou. The adaptive lasso and its oracle properties. *Journal of the American statistical association*, 101(476):1418–1429, 2006.

[166] Hui Zou and Trevor Hastie. Regularization and variable selection via the elastic net. *Journal of the royal statistical society: series B (statistical methodology)*, 67(2):301–320, 2005.

[167] James Zou, Christoph Lippert, David Heckerman, Martin Aryee, and Jennifer Listgarten. Epigenome-wide association studies without the need for cell-type composition. *Nature methods*, 11(3):309, 2014.

[168] Bruno D Zumbo and Donald W Zimmerman. Is the selection of statistical methods governed by level of measurement? *Canadian Psychology/Psychologie canadienne*, 34(4):390, 1993.

Index